ISLINGTON

Please return this item on or before the last date stamped below or you may be liable to overdue charges. To renew an item call the number below, or access the online catalogue at www.islington.gov.uk/libraries. You will need your library membership number and PIN number.

1/19

1 5 MAY 2019

D0765941

Islington Libraries

020 7527 6900   **www.islington.gov.uk/libraries**

Alex Beard has worked in education for a decade. After starting out as an English teacher in a London comprehensive, he joined Teach For All, a growing network of independent organisations working to transform education globally. He is fortunate to spend his time travelling the world in search of the practices that will shape the future of learning, and has written about his experiences for the *Guardian*, *Financial Times*, *Independent*, *Evening Standard* and *Wired*.

———————

'Readers should accompany Beard on his travels – not only because the subject of his inquiry is so important, but because after myriad engaging anecdotes and encounters, he arrives at some conclusions with universal relevance'   Miranda Green, *Financial Times*

'Invigorating . . . one of the most optimistic, thought-provoking and ambitious educational books of recent years. *Natural Born Learners* is audacious, sassy, unafraid of big questions about what our children deserve and what our culture needs from education ... bold and exuberant'

Geoff Barton, general secretary of the Association of School and College Leaders, *TES*

'In this vital and vitalising exploration of the future of education, Alex Beard has written a compelling work of non-fiction. A meditation, a rallying cry and an inspirational call to cut a new educational path, this is an incredibly important book. As Beard powerfully illustrates, if we get education right for everyone then we'll be well on the way to solving so many of our other problems'

Owen Sheers, *Metro*

'A timely and passionate book on the need for a global educational renaissance'   Jane Shilling, *Mail on Sunday*

'Wide-ranging, often humorous and consistently fascinating, this is a book for all those interested in learning – a process that, as the author stresses, should be lifelong'   Stephanie Cross, *The Lady*

Alex Beard

Natural Born Learners

Our Incredible Capacity
to Learn and How We
Can Harness It

W&N
WEIDENFELD & NICOLSON

For Daisy

First published in Great Britain in 2018 by Weidenfeld & Nicolson
an imprint of The Orion Publishing Group Ltd
Carmelite House, 50 Victoria Embankment
London EC4Y 0DZ

An Hachette UK Company

1 3 5 7 9 10 8 6 4 2

A CIP catalogue record for this book is available from the British Library.

ISBN (paperback) 978 1 4746 0473 4
ISBN (ebook) 978 1 4746 0474 1

Typeset by Input Data Services Ltd, Somerset

Printed and bound by CPI Group (UK) Ltd, Croydon, CR0 4YY

www.orionbooks.co.uk

Life should be an unending education; everything must be learned, from talking to dying.

Gustave Flaubert[1]

CONTENTS

A Tale of Two Academies

The future is here, it's just unevenly distributed.

William Gibson

Into the Light

The Hippeios Colonus lay a mile north of ancient Athens. A hill thick with vine, olive and laurel, topped with a temple to Poseidon and a sacred grove to the Eumenides, or Furies, it was said to be the resting place of Oedipus and birthplace of the great playwright Sophocles. Climbing it in 385 BC on a warm Mediterranean evening, the pink light playing on the Aegean and the scent of oregano wafting up from the baking ground, you might have looked down to the west and caught a glimpse of one of the great inventions in human history – the school.

This school was situated in the groves of Akademeia, named after the Greek hero Akademos, one-time saviour of Athens. The area was known at the time as the destination for an atmospheric ritual, a torch-lit race from the city in which late-night runners sped along a path flanked with the graves of dead Athenians to arrive first with their flame at the altar of Prometheus. It was also sacred to Athena herself, goddess of wisdom.[1] Back then a middle-aged and much-travelled nobleman called Plato had just taken over a large part of it to host his new 'Academic' club, devoted to the pursuit of knowledge.

Plato's Academy quickly became an intellectual hothouse, the MIT or Cambridge University of its day. The many illustrious alumni included Aristotle, who would go on to found his own Lyceum, and tutor Alexander the Great. Its ideas would inform civilisations. Learning consisted of discussion around particular texts and case studies, with a teacher – often Plato himself – posing problems for the members to consider in conversation, just as they do at Harvard Business School today. Topics included mathematics and philosophy and ranged from the scientific analysis of the movements of heavenly bodies to consideration of the best modes of government. Plato wrote down many of these cases in *The Republic*, interposing his former mentor Socrates as the textual teacher and other members of his entourage as his protégés. One of the most famous is the Allegory of the Cave.[2]

Picture an odd grotto. In it a group of people are chained to the floor on one side, their legs and necks fixed in such a way that they can only stare at a blank wall. They have been secured in that position since early childhood and the cave wall in front of them is all they have ever seen. Behind them is a fire and in front of it a low screen over which a variety of objects are raised by hidden helpers, who also make sounds. The prisoners see moving shadows cast upon the wall and attribute the noises to the spectral shapes. They assign meaning to what they see and start to understand the play of shadows as reality. It is all they know. The flickering forms and sinister sounds are their whole world.

What would happen, asks Socrates, if one of these prisoners were freed?

On turning, the bright light of the fire would dazzle the prisoner at first and he would be unable to make out the shapes of objects, or make sense of the new visions assailing his senses. He would turn fearfully back to the wall. Imagine then that someone dragged him forcibly from the cave, past the fire and into the sunlit lands above ground. The prisoner would be angry, resistant and near blind, holding tightly to his old ideas of existence. After the wall and the shadows, the technicolour world about him would seem a shocking hallucination. But slowly his eyes would become accustomed to the light, the pain would recede and he would come to terms with a new and infinitely more beautiful reality. He would bless himself for the change and rush to free his other cave dwellers.

We can read in Plato's Cave a parable about learning. Two and a half thousand years later its imagery of light and dark as metaphors of knowing and ignorance remain familiar. We see ourselves as sentient, conscious and rational beings, as people who have 'seen the light'. We sense that more and more of humanity has passed out of the cave and into the world above. For Plato, however, it was clear that most of us were yet to make that journey. His mission – and that of his school – was to lead more people into the sunlit lands of enlightenment. It was the work of the philosopher to push at the frontiers of human apprehension, to more fully understand the world, and to better decide how individuals and societies should live in it.

Today we face an even greater challenge. We believe that *everyone* should be well educated, not just rich noblemen. Our young must thrive in a world whose pace of change seems destined to increase exponentially, whose future is unclear. Yet when we look at our school systems, we don't see Plato's clear-sighted mission of human betterment, but ailing bureaucracies struggling to maintain bright points of light amid the gloom. Like stars on a hazy night above modern Athens.

If at First You Don't Succeed

I could, so I taught. One bright September morning a decade ago, I pedalled my way down the Old Kent Road – London's cheapest Monopoly property – to begin my life as an English teacher. The school was in one of London's poorest and most diverse neighbourhoods, Elephant and Castle, named after an eighteenth-century coaching inn. I knew it only by reputation. The area was dominated by two housing estates, the Aylesbury and the Heygate, whose maze-like walkways and dingy stairwells meant they were no-go zones after dark.[3] Walworth School didn't seem a lot better. At a meeting of new teachers in the area, a veteran from a nearby secondary revealed: 'That's where we tell our kids they'll end up if they don't behave.' My first assembly would begin with the stark announcement that a 14-year-old boy had died, stabbed after a game of football.

That first day marked a new dawn for Walworth, however.* It had just begun life as an academy, part of a government scheme to give more money and more autonomy to struggling inner-city schools, though it was a far cry from the Athenian original. I pulled up to the gates that morning with the potent mix of nerve-shredding trepidation, rank incompetence and platonic ideals familiar to all beginning teachers. I was scared, I was ill-prepared, but I knew – *knew!* – that I'd be *Dead Poets Society*'s Robin Williams by the end of my third week. After all, everything I'd learned in my own school and university days told me that education was simple. You posed the right problems, outlined interesting thought experiments, then sat back, engaged minds and discussed.

My first classes were failures. Initially placed in the lower school, I found the younger kids enthusiastic but unfathomable. Eleven-year-old Kai approached me during one class in his socks claiming his shoes had *dropped* out of the third-floor window and asking if he could retrieve them. A particularly wild break-time ended with Shaun at war from one side of the class with Marcel on the other, using chairs for ordnance and tables for cover. Every lesson began with a dispiriting chorus of lost books, missing homework and attempts to break out to the hallowed freedom of the toilet. News of my struggles spread and competent fellow teachers were regularly called in to assist. The groves of Academe seemed a distant dream.

Over in the upper school, things didn't go much better. Although intermittent attempts at reading and writing were known to break out among them, the habits of the older kids more typically comprised a mix of uninterested gazing out of the window, mind-boggling feats of misunderstanding and an unending ability to find new variations on 'Your mum' cusses. One Year 10 class was like a UN General Assembly, comprising 30 kids of British, Irish, Chinese,

---

* Walworth Academy has changed in the past decade. In 2008, it came in the bottom 12 per cent of schools in the UK for students achieving five A*–Cs including English & Maths. Today it is rated a good school, with kids performing around the national average. For disadvantaged students, it ranks in the top third nationally for attainment, and the top 20 per cent for progress.

Jamaican, Liberian, Congolese, Afghan, Somali, Sudanese, Nigerian, Turkish, Portuguese and Vietnamese backgrounds, with just as many disagreements. Many spoke no English at home. But to me their situation was increasingly unfunny. They had an English GCSE to prepare for, a high-stakes exam that would decide their future. The class was averaging Ds, Es and Fs. They'd have to score As, Bs and Cs within eighteen months if they were going to stand a chance.

As part of the course, over the next year and a half together we would cover two Shakespeare plays for coursework. I was looking forward to this as an opportunity to flex my intellectual muscles, whatever the doubters said. (In my interview for the teaching post I'd mentioned my love of literature, and plans over the summer to fill in dangerous gaps in my subject knowledge by reading classics – Milton, Marvell, Woolf and Eliot. The interviewer had looked at me patiently and replied, 'That's great, but I'd start with *Holes* and *The Boy in the Striped Pyjamas*.') The school had chosen *Romeo and Juliet* and *Macbeth* as set texts, and I spent a few weekends sharpening my opinions. The kids would excel, despite their challenges. This is what I had entered teaching to do.

Things did not go well. Progress through *Romeo and Juliet* was stultifyingly slow. We spent a week trying to understand the prologue and ultimately read only a few scenes of the play. The kids could express opinions about it if I created fill-in-the-blanks activities with a choice of three adjectives, but were otherwise short of ideas. After we watched the Baz Luhrmann movie adaptation to fill in the story gaps, every subsequent essay on the sixteenth-century text featured personalised revolvers, Miami muscle cars and exploding petrol stations. Despite my high hopes that the class would find fulfilment in the pursuit of understanding and a love of literature, in their first coursework essay they remained resolutely stuck on low grades.

I thought back to my own school days. I'd had the good fortune to attend a good primary school in a small Midlands town, where the teachers, who were more like surrogate mums or grannies, had inspired me. From there I'd gone on a scholarship to a private school. It boasted the largest single expanse of mowed grass in the UK and an altarpiece – referred to in Old French as a *reredos* – reputedly valued at £6,000,000 (it was the chaplain's favourite game during RE lessons to invite boys to try to evade the security system's laser beams to reach the altar, like trainee gentleman cat burglars). We called our teachers dons and head teacher the Warden. In our English classes, we'd pontificate on Jane Austen and T. S. Eliot, and then all write A or A* essays. There was a sense of inevitability about it – just as there was a sense of inevitability about the failure of the class that I was teaching.

Yet a realisation drove me on. Getting to know these South London kids, working with them and talking to them each day, I quickly found that there was no fundamental difference between them and the kids I'd been at school with. They had the same dreams, the same camaraderie, the same feuds and the same teenage angst. Their parents, like mine, desperately wanted them to do well, and to be happy. They didn't wear gowns to school or have access to a pack of beagles, but these were superficial details. In potential, in ambition and especially in jokes, they were more than a match for my privileged peers at boarding school. But where society had given us a leg-up, it was letting these kids down.

It all felt a long way from Plato. A long way even from Robin Williams. The class was failing, and I was at a loss.

This Solves Everything

This book is inspired by those kids at that South London academy. As a teacher I was startled to realise that school is fundamentally the same now as it was in Plato's day. A time-travelling child from ancient Athens might be mystified by our smartphones, overwhelmed by our populous cities and alarmed by the cars on our roads. But she would have no trouble at all recognising a classroom with its teacher and pupils. With all the progress we've made in other fields of human endeavour – in medicine or neurology, psychology or technology – aren't we long overdue a revolution in the way that we learn?

The 2,400 years that separate us have witnessed epochal changes: near-incomprehensible growth in the global population; huge agricultural, industrial and technological revolutions; incredible transformations in the ways that we create and disseminate knowledge; new forms of social and political organisation; insight into the secrets of the mind. These have thrown up the myriad challenges of globalisation, automation and climate change. If we're to overcome them, we must further increase our extraordinary ingenuity, more fully develop our skills and radically improve our co-operation *as a species* to unleash our full human potential. Learning must be the cause of our generation.

How should we approach education today? Over the pages that follow, I'm going to take you on an exploration of what it means to thrive in today's rapidly changing world and what we can do to ensure that all of our kids do. In Plato's era the main concern was to push at the frontier of human knowledge and understanding for grown men. Today, while we continue that quest, a more important question is how we extend access to the furthest reaches of human development to every child, to all people. Our aim is no longer the flourishing of a few philosopher-citizens in an ancient city-state, but the flourishing of a philosopher-race to steward our high-tech, globalised civilisation.

Following those first faltering classes a decade ago, I confronted my bafflement at the low learning levels of my pupils and embarked upon a lifelong quest for new ideas and exciting innovations that could inspire us to reimagine schools and remodel the creaking ziggurats of our global education systems. In this journey, time and again I asked myself the question central to all learning – why? Why do schools now look so similar to schools in ancient Athens? Why do we prize academic success above all? Why are kids so often unhappy in their learning? Why do we continue to pursue an industrial model that businesses have left behind? And throughout it all I have obsessed over a single goal: to show what learning in the twenty-first century *should* look like.

The search you're about to join me on has taken me across the world, from the intelligent machines of Silicon Valley to the exam factories of Seoul; from Finland's greatest teacher to Britain's brightest student; from the MIT professor raising a robot to the Hong Kong schoolkid battling a superpower; from teachers trained like athletes to students learning without teachers. I have visited schools on every continent of the earth, talked to the leading neuroscientists and experimental psychologists, met the most fabled educators. I have explored the frontiers of the mind and the limits of the latest technology. I even ended up in Hollywood. The good news, as you'll find out, is that everywhere I have seen signs that we are on the cusp of a revolution in the way that we learn.

This book outlines for everyone the three key tenets that will drive that transformation. The text is arranged into three parts.

The first argues that we must *think anew*. Science has begun to delve deep into the inner workings of our brains, showing that each of us has a far greater capacity for learning than we realise. We're literally *natural born learners* – every one of us – but we're too often held back by the false belief that our intelligence is fixed. It isn't. Our understanding of the mind is limited by metaphors of computing, but it is not a machine to be programmed by schools. The brain is alive, unruly, engaged in an unending process of inquiry. As medicine underwent a scientific revolution in the nineteenth century, so can education today. Thinking anew about human development focuses our attention on upgrading ourselves, not our technology.

The second part urges us to *do better*. Our schools are reasonably effective at achieving what they set out to: producing a solid blue and white collar workforce well drilled in what Sir William Curtis dubbed the 'Three Rs' of 'reading, 'riting and 'rithmetic' in an 1825 speech in the British Houses of Parliament.[4] But as automation and globalisation cause traditional jobs to disappear, so must traditional models of schooling give way to those that grow creativity and purpose. A craftsperson aspires to make works of great beauty, is an able user of the most appropriate tools and feels flow when mastering skills. Doing better means beginning with human creativity. We must ensure kids develop the means to express themselves and find a place in the world. This applied scholarship is our noblest aim.

The final part explains why we must *take care*. The education of our children is a perpetual labour, and it remains the most important undertaking of our race. Yet in recent years it has lost touch with its innate human imprint. Schools borrow increasingly from paradigms of the factory or market, vaunting efficiency and competition. This has brought great surges in literacy and improved exam results, but it has pitted kids against each other in a brutal race to the top, narrowed the parameters of learning and taken economic output as education's only measure. In the future, we must rediscover the ethical and human dimensions of learning. Taking care requires us to build our systems around shared values, not new technologies, framing them as ecosystems rather than corporations. The well-being of our species, and of our planet, depends on developing our social and emotional intelligence. We must learn to co-operate in building the future we wish to see.

We begin our endeavour from a stable foundation. There is no better time to be a pupil than today, with more than 1.2 billion children in school. Standing in front of them in classrooms from Lima to Lucknow are more than 50 million teachers, almost all passionate, able and committed.[5] And yet unless we can rapidly adapt the way children learn to the evolving needs of the world today, we risk a lost generation. Six hundred million of those kids are currently failing to master the basics, let alone the tools they need to succeed tomorrow.[6] Meanwhile, our experience ties us to the past. School is something that everyone feels expert in. Most of us have spent at least twelve years – more than the fabled 10,000 hours – in classrooms. But we've learned the wrong lessons. Not quite an art and not yet a science, the field of learning still paradoxically seems at times devoid of a deep, unitary expertise.

The time for us to unite that expertise is now. Through thinking anew, doing better and taking care, we can bring about a twenty-first-century enlightenment in education that ensures more and more kids fulfil their potential. As the physicists have their Theory of Everything, and the philosophers their Absolute Mind, so we educators must strive for the flourishing of all humanity. As *Homo sapiens*, wisdom – learning – is our defining characteristic, marking us out from our hominid ancestors. The cultivation of this attribute should be our species' highest purpose. We must use this moment of technological disruption, with its jobless future, diminishing resources and driverless cars, to step back and imagine a world that places human development at its centre. Everything depends on our ability to do so.

Failing Again, Failing Better

A year later and my GCSE class was about to graduate into Year 11. Their first exams were a little over six months away. Soon they would be leaving school for sixth form colleges and universities beyond. Their grades would be the only evidence of their abilities they could show to future institutions or employers. The other path, with rates of youth unemployment above 50 per cent for those without degrees, did not bear thinking about. Dreams remained in the balance.

One day we confronted a particularly difficult passage in *Macbeth*. Fabrice, a 15-year-old boy who had been born in the Congo and come to London via some years in Rotterdam, was wrestling with an idea. He'd been leader among a group of troublemakers that were finally developing the abilities to enjoy learning. The topic was stage directions, and we were reflecting on the decisions that a theatre director might make about performing the 'Is this a dagger which I see before me . . . ?' scene. The question I had posed involved a difficult combination of higher-order thinking skills – what would the implications be for the audience's understanding of Macbeth's character if the director chose to show, or not to show, the dagger?

It was a problem worthy of Plato's Cave. Amir, a slight teenage boy from Afghanistan who had been moved to my class as a trouble-maker, had his hand up desperate to answer. Triggered by his fresh cultural perspectives – he believed wholeheartedly in the evil magic of cats – he had become entranced by the themes of witchcraft and sorcery in the play, and had used these as a foothold to develop striking and original opinions on Shakespeare. Fabrice continued to consider.

'*Oh*,' he exclaimed suddenly, 'I get it.'

As Amir bounced on his seat with his hand raised, Fabrice carried on confidently. It was one of the few times that I remember seeing *visible* learning in my classroom. I could almost hear his brain whirring. He was mastering a new and complex way of thinking.

'If we see the dagger, then we might think the witches have used magic to trick him.'

'And if we don't see the dagger?' I replied.

'If we don't see the dagger, then we would think . . .' He pondered. Amir continued to bounce. Suddenly Fabrice's face flickered with understanding: 'We would think that he is *bare* crazy!'

He turned round to Amir and held his finger to his lips, like a footballer scoring against an arch rival.

It was a watershed moment. Fabrice went on to score As and Bs in his GCSE coursework – Amir got As and A*s. When the exams came around, almost all of the class achieved at least the C that they required to continue on to higher education.

I was elated. But I was also unsatisfied. The kids had succeeded, but only in a narrow sense. They achieved their necessary GCSE grades, but school hadn't been able to offer them much else. With a proper go at it – and a better teacher – they could have excelled. They were far behind when they came to my class, with many unable to read or write properly after eight years of education, and their Cs, though cherished, didn't suggest they were ready to change the world. On top of that, I wasn't entirely sure how we'd made progress. There had been a lot of sweat – and tears – to my approach, but no science. My early incompetence and lack of imagination had cost us valuable learning time. Surely, given all that we knew these days about the mind, the brain, the body, about human behaviour and the science of performance, we could devise a better approach than this? Surely, given the challenges our society faced, we had to?

The small success I experienced with my Year 10s gave me confidence in the power of education in the twenty-first century to fuel the lives of individuals and to power societies. But we'd have to get it right. We'd have to think anew about the potential of our kids, better equip them to use the tools of today and ensure we took care of them all. Every child was born to learn, but our systems, rather than building on that potential, seemed to be inhibiting it. I set off that day on my journey. Starting out in Silicon Valley, I'd travel across new countries, through new roles and into new classrooms, to find out how we might get started.

If these kids could succeed – from a backward starting point, with a new teacher – then all kids could. The trick in this complex, ever-shifting and rapidly changing world was to ensure that all of them did.

PART I

THINKING ANEW

CHAPTER 1

Artificial Intelligence

Beware Geeks Bearing Gifts

> Whom the gods would destroy, they first
> call promising.
>
> Cyril Connolly

The Robot Teachers are Coming!

Brett Schilke sat in a back room of Singularity University's Mountain View headquarters talking about the future. Since his school days he'd been on a mission to revolutionise learning. 'I was *that* kid,' he said, 'I was like, "Tell me why I have to learn this." I had one teacher who just pissed me off. He had the same answer every time I asked – it might be a question on *Who Wants to be a Millionaire?* And I was like, you literally can't come up with anything better than that? Can I *leave*?'

Schilke had worked in education since he graduated from college, where naturally he excelled. He was an unashamed enthusiast: an adventurer, educator and instigator who – *Welcome to California!* – loved stories, puns and high-fives. After starting out running cultural, arts and education development programmes in Siberia and Transylvania – 'Yes, the former gets cold. No, the latter does not have vampires'[1] – he had returned a few years earlier to the Midwest to run IDEAco, an education non-profit whose projects included City X, a problem-solving and 3D-printing curriculum for kids. He then joined Singularity University, the organisation set up by the high priest of futurology and author of *The Singularity is Near*, Ray Kurzweil, 'to educate, inspire and empower leaders to apply exponential technologies to address humanity's grand challenges'.[2]

The Singularity was Kurzweil's term for a hypothetical point in the future when artificial intelligence would become trillions of times more powerful than our human minds, ushering in a new civilisation 'that will allow us to transcend the limitations of our biological bodies and brains' by merging with our technology.[3] A cool notion, if a little scary. While Kurzweil, who now leads Google's AI division, foresaw a utopia in which our augmented minds achieved unimaginable feats of cognition, others envisioned a human face crushed for ever under a hyper-intelligent robot boot. Singularity University – SU to believers – could be interpreted as Kurzweil's effort to tip the outcome towards the former.

Brett Schilke had recently been appointed SU's director of Youth and Educator Engagement. It was his job to be obsessed with the future – he was careful to distinguish it from education or school – of learning.*

Behind him hung a painting of a robot on a Harley-Davidson leaping from a tower of iced doughnuts towards a golden horizon. He spoke fast, ideas puttering up like popcorn in the microwave.

'It's a super-exciting time to, like, be alive,' he said. 'That sounds really corny, but it *is*. It's awesome. It's just so unexpected what you can do every day.'

He looked at me with clear eyes.

'It's *wild*.'

He was talking about how technology was changing the world,

---

* Since at least Aristotle, we've been drawing a distinction between what happens within the four walls of a classroom and an ideal form of learning. For instance, Mark Twain is supposed to have written that you shouldn't 'let your boy's schooling interfere with his education'. Even the World Bank's 2017 World Development Report (which was completely dedicated to education for the first time ever) now opens starkly with the idea that 'Schooling is not the same as learning'. I think there's something in this, as you can tell from the title of my book.

and how the world – and our schools – had to change with it. In Silicon Valley the idea that we humans are capable of more was as commonplace as the belief that technology is a purely positive force. Schilke had drunk that Kool Aid. In fact, he added that we must *learn* and *create* together to achieve our machine-assisted potential. For historical reasons, we were not yet doing this.

'We have a system that was designed for the Industrial Revolution. That's where modern education came from. We needed to produce a massive workforce that does simple tasks over and over and over and over. And how do you do that? Well, let's get them when they're young and teach them to sit up straight and raise their hand.'

He paused, a little hysteria in his voice.

'It's all about building this almost *militarised* group of people.'

This was broadly true. Education systems had been influenced by a militaristic model. In the 1830s Horace Mann, then education secretary in Massachusetts, pioneered a state-wide form of schooling that became the basis for free and universal education in the US. The model was inspired by a visit he had taken to Prussia, a country renowned for its strict hierarchies, obedience to power and military might, where a few decades earlier Frederick the Great had signed into law the world's first national system of education. That paradigm, strengthened by the ideas of industrialisation, mechanisation and massification, came to define universal-education systems that soon cropped up all over the world. But thanks to computers and other new technologies, Schilke felt that finally these notions were being challenged.

I'd made Silicon Valley the first stop on my journey into the global learning revolution so that I could find out how. The techno-humanists of the Bay Area exerted a powerful influence over our view of the future. And I wanted to know what artificial intelligence could tell us about the power of our own minds. Was human learning becoming obsolete, as some suggested, or could we use computers to augment our brainpower to unimaginable levels? I thought we risked underestimating our own natural capabilities, fittingly adapted over millions of years of evolution, and instead had to think anew about our own capacity for learning in the digital age. If we could better understand our brains, and learn to use our technology wisely, the potential might be much greater than we realised.

The first thing to understand, thought Schilke, was that we had not simply to invest in the latest gadgets, but to radically change how we thought about learning.

'SU focuses more on how you equip teachers for the larger technological and social trends that are coming,' he explained. 'We don't teach 3D printing to teach 3D printing, as a job skill like you learn to be an accountant. We teach 3D printing to teach 3D *thinking*, to learn how you conceptualise ideas.'

This focus on higher-order thinking was increasingly backed up by research. Two futurist economists at the Oxford Martin School, a centre set up to predict and plan for societal changes to come – had concluded a couple of years previously that of the 702 current jobs done by humans (by their calculations), around half might soon be taken over by artificially intelligent machines.[4] If during industrialisation the robots had eaten muscle jobs, in the era of computerisation they were coming for those of the mind. This posed a double challenge to schools, first to incorporate the newest technologies in the learning process and second to reimagine the content of a useful education. If anyone in the world knew how to meet these challenges, I thought it would be technophiles like Schilke.

Earlier we walked round the campus where SU had its home, an old NASA research institute and military base dominated by a huge skeletal structure, the uncovered frame of the old hangar where airships were constructed in the 1950s. Now empty, it was sometimes used by Google as a venue for exclusive staff parties – their campus swallowed the land on three sides. Just beyond the fence we could see the Moffett Field airstrip where the tech giant was testing pilotless flying transports, and which was used by President Obama to land Airforce One on his visits to the Bay. An eagle circled overhead, nature's proto-drone.

Schilke revelled in SU's place at the heart of all this innovation. He pointed out the base's dilapidated McDonald's, long since repurposed. 'In there is a project to map the surface of the moon,' he said, 'it's *so* cool. They call it McMoon.' On site was a *Who's Who* of tech companies: Tesla, Carnegie Mellon, Moon Express. In the distance were the looming towers of NASA's rocket-engine testing facility and dotted around the car park the latest hybrid vehicles and electric cars. This sun-kissed place, with its mountain backdrop and hulking government warehouses, now commandeered by friendly-faced tech corporations, was at the heart of all the new in the world. It was intoxicating.

The final stop on the tour was to be the Classroom. Schilke talked excitedly of the toys that we'd find there. When I'd been teaching Fabrice and Amir, the latest technology meant beaten-up laptops that had got in the way of progress. Now, I couldn't wait to see the gadgets, Virtual Reality lecture halls and robot teachers, 3D printers and nano-materials. I thought about Neo in *The Matrix* downloading learning into his brain in seconds. Perhaps technology really was on the cusp of revolutionising learning. Maybe, in that room, was the future of school.

I'd had my first sip. The Kool Aid didn't taste too bad.

Are Computers the New Books or the New Televisions?

For people who deal exclusively in preparing others for the future, we educators are surprisingly reluctant to embrace the new. Our own experience biases us against it. Wasn't school just so for us, and didn't we turn out alright? Certainly at St James' Primary School in the 1980s there was not a single computer. My Year 1 teacher – Mrs Calcutt – outlined our first words and numbers in chalk (which we only occasionally used as a projectile). The tools of learning were pencils, paper and books. We practised handwriting and met the inhabitants of *Letterland* from Annie Apple to Zig Zag Zebra. It was decidedly no-tech. And if that worked for us, we now tell ourselves, it will work for our kids.

It's wise to be somewhat circumspect about the potential of the latest technology to change the way we learn. The lustre of the new has a tendency to hypnotise. In 1922 Thomas Edison predicted a dramatic transformation in public schools:

I believe that the motion picture is destined to revolutionize our educational system ... in a few years it will supplant largely, if not entirely, the use of textbooks. I should say that on the average we get only about two percent efficiency out of textbooks as they are written today ... Through the medium of the motion picture ... it should be possible to obtain one hundred percent efficiency.[5]

The trend continued. In 1966, dazzled by the potency of advertising in shaping the habits and behaviours of the American people, President Lyndon Johnson was moved to intone that 'unhappily, the world has only a fraction of the teachers it needs', but that this could be compensated for by 'educational television'.[6] Unless I've just not been to the right classrooms, neither of these revolutions came to pass.

Yet new technologies *have* at times radically transformed learning. Five thousand years ago, the invention of writing enabled humans to transfer knowledge through space and time, storing it outside of our minds as never before. Even then there were sceptics, with Socrates lamenting the written word in the *Phaedrus*, arguing that it undermined our capacity for memory and distanced us from authentic truth.* But the transformation effected was in no doubt. No longer would learning be defined by the quality of the tutor you could afford; nor was the evolution of knowledge limited to a dialogue between two people. Now across space and time ideas could be shared and adapted through the minds of the many, and new structures of thought created. This transformation was boosted a little over 500 years ago when the printing press and first vernacular Bibles precipitated a tipping point in the access to knowledge of the masses. The availability of cheap, plentiful books played a huge part in the great surge of literacy experienced by the West in the late nineteenth century.

It looked like our parents were right – books *were* better for us than TV. So, if we were unsure about the likely impact of technology

---

* In Plato's *Phaedrus*, the Egyptian god Thoth offers King Thamus writing as a 'remedy' (*pharmakon*) that can help memory. Thamus is doubtful of the tonic effects of the invention, suggesting that on the contrary, once people begin to write things down, they will no longer use their memories, but refer always to what is written. Among others, French philosopher Jacques Derrida has written brilliantly on the paradoxical idea of the pharmakon, the cure that is also a poison.

in education today, the question we had to ask ourselves was this: were computers the new books or the new televisions?

Why Computers Might be Books Squared

One measure of a person's education is their intellect, and the cerebral world of chess has long been its proving ground. The Cold War showdown between Boris Spassky and Bobby Fischer in 1972 captured the imagination of the world precisely because it could be construed as a victory for the American over the Soviet mind (no matter that Fischer was the son of European immigrants). While the young maverick and the old master were squaring off in the match of the century in Reykjavik, around the same time computer scientists in the US were working on a seemingly more innocuous conundrum – could a computer beat a person at chess?

By 1972 there was already pretty strong evidence that the answer was yes, at least at the amateur level. In 1967 a group of MIT students put together a computer called Mac Hack IV to take on the philosophy professor Dr Hubert Dreyfus in a game. A strong amateur player and leading human mind, he looked down on the gimmicky machine, declaring that no computer could yet beat even a ten-year-old child at chess.[7] From a winning position Dr Dreyfus's fallibility got the better of him, and he lost to the machine. The same year, Mac Hack IV became the first computer to win an official tournament match. Over time, these challenges became the battleground of human versus machine mind. The most sought-after scalp was to be hard won – that of the world's leading grandmaster.

In 1997, after decades of attempts, a team at IBM felt that they had finally prepared a machine that was up to the task. Echoing the 1972 match of the century, Deep Blue (whose forebear Deep Thought had been named after the all-knowing computer mind in Douglas Adams's *The Hitchhiker's Guide to the Galaxy*), the first thinking machine, was to take on humanity's leading player, Gary Kasparov, who a year earlier had defeated a similar machine in Philadelphia. The rematch took place in New York, where, supported by a team of software developers (who were later accused of assisting the computer in ways that went against the rules),[8] Deep Blue defeated the grandmaster 3½–2½ in a tense and controversial match.[9]

Robots 1 Humans 0.

But although it was a big deal – machines could think! – it wasn't *that* big a deal – they could only think like machines. As Kasparov later wrote, 'Deep Blue was only intelligent the way your programmable alarm clock is intelligent. Not that losing to a $10 million alarm clock made me feel any better.'[10] Also, chess was a pretty niche pursuit. The point of human intellect wasn't only to win chess matches, and schools weren't tasked only with producing grandmasters.

But IBM, perhaps forgetting whose side they were on, didn't stop there. They apparently took a special pleasure in sticking one in the eye of humanity. Looking for a new challenge after the Kasparov match – so long loser! – they landed on the US game show *Jeopardy!*. Here the machine would have to demonstrate a much more human set of skills, such as the acquisition of a lot of useless pub-quiz-type knowledge, and the ability to interpret the puns and wordplay that were crucial to the game. The developers set about creating a machine that would be able to think more like a human and – giving our likely future overlords a misleadingly avuncular sidekick quality – named it Watson.

In 2011, in a ratings-smashing televised spectacular, Watson defeated the all-time greatest human *Jeopardy!* champions, Brad Rutter and Ken Jennings. It wasn't even close. The game ended with Jennings on $24,000, Rutter on $21,600, and Watson on $77,147, meaning IBM scooped (and donated to charity) the $1,000,000 prize. Afterwards Jennings wrote that 'just as factory jobs were eliminated in the twentieth century by new assembly-line robots, Brad and I were the first knowledge-industry workers put out of work by the new generation of "thinking" machines', adding that '"quiz show contestant" may be the first job made redundant by Watson, but I'm sure it won't be the last'.[11]

While books had facilitated a paradigm shift in the way that knowledge was codified, stored and shared, they did not appear to *think*. But over the past 50 years, it has become clear that computers can use, apply and even generate knowledge (which sounds suspiciously like the marking criteria for GCSE English). Deep Blue showed remarkable tactical acuity to defeat Kasparov, looking into millions of positions and making seemingly creative moves to wrong-foot the grandmaster. Watson mastered punning and wordplay and had in its memory a reservoir of over 200 million pages of recondite facts. Even if they were only artificially intelligent in a narrow sense (neither, for example, could tell even a simple joke), the breadth of the computer mind was clearly growing. They were, on some level, *thinking* machines.

If Brett Schilke was excited by the potential of artificial intelligence for upgrading the way we learned, he wasn't alone. It seemed to me that if writing and books had revolutionised human cognitive development, then computers were about to do the same. In Silicon Valley, one of the first schools to invest heavily in our new computer-teacher overlords was – as the drone flies – just 20 miles from SU. I'd heard that they were getting computers to do the work of teachers, and wondered what that meant for the minds of our kids. I decided to pay a visit.

The Teaching and Learning Machines

It was a bright October morning and as the workers of Silicon Valley collected their drive-thru Starbucks, the 400 students of Rocketship Fuerza Community Prep were filing out of the school yard. They had just completed 'The Launchpad', a daily routine in which the mic'd-up head teacher, Ms Guerrero, readied the young Rocketeers for class, leading them through the pledge of allegiance, whole school cheers, songs and the handing out of prizes for things like 'grit' and '*ganas*'.* The highlight had been a singalong to Des'ree's 'You Gotta Be' and a whole school dance routine – parents included – to 'Shake It Off' by Taylor Swift.

'It's morning coffee for the kids,' said a teacher. It looked like it.

---

* *Ganas* is a Spanish word that can loosely be translated as 'guts'. It was popularised throughout US schools largely thanks to the movie *Stand and Deliver*, a true-life tale of the fabled maths teacher Jaime Escalante, who came from Bolivia to East LA and taught AP Calculus to kids from the wrong side of the tracks, leading them – hey, they made a movie about the guy – to success against the odds. 'Students will rise to the level of expectations,' he says in a pivotal speech, '*ganas*, that's all we need, *ganas*.'

The Rocketeers leaving in teams – The Broncos! The Spartans! – were pumped.

The strange-sounding terminology was carefully chosen. Rocketship lifted off in 2007 as the first in a new wave of West Coast schools that would self-consciously surf the tech tsunami. Software entrepreneur John Danner was one half of the founding team. He saw an opportunity to harness the growing potential of machine learning to personalise the school experience for each child. The zero cost replicability of digital tools also appealed to his entrepreneurial nature. They would rapidly test and scale a hyper-efficient school model that within 20 years would reach 2.5 million kids in 2,500 schools nationwide. If AI could win *Jeopardy!* it could teach a few elementary schoolkids how to solve maths problems.

The school's other co-founder was Preston Smith, a career educator and teaching superstar, who'd run highly successful schools for marginalised kids in the San José area. When I met him in his office downtown he explained how thinking machines were beginning to help schools. 'There's a place for technology around instructing things that are really hard for teachers to teach. I think math, how you can visually do things is profound. Practice is profound. Getting things off teachers' plates because they are really way too talented to be doing sounds and letters with all their kids. We think about the opportunity in terms of time. It's gonna help my teacher not to have to teach this. It's gonna help my teacher be more effective. It's gonna help my Rocketeer master this standard more quickly. It's gonna buy back time to do more critical thinking, more higher-level things. That's what we obsess over.'

Rocketship was making a big bet on the ability of technology, and particularly AI, to automate some learning experiences.

Underpinning this approach was the Learning Lab, a place where kids would go each day to be tutored by intelligent machines.

After the kids had finished their breakfast, Ms Guerrero and I headed over. The Lab was a cavernous 2,000-square-foot room with whiteboards on either side, a school hall X.0. In the centre two adult supervisors sat behind a circle of desks. Arrayed either side of them in six long rows facing out towards the whiteboards were 100 five-year-olds. All wore the distinctive purple uniform of Rocketship and all had a laptop in front of them and a set of outsize headphones over their ears, like miniature novices at a space-age seminary. Half were working on ST Math, an online arithmetic platform, and half on a reading program called Lexile. They were busily completing their problems, heedless of me, a looming six-foot-four visitor.

Save for the soft tapping of small fingers on keyboards the room was eerily silent.

I crouched down to see what one young girl was up to. Her name was Martha and she was playing a computer game – it would have looked hi-tech in the early Nineties – navigating a rudimentary space shuttle through a field of asteroids, with limited success. I pointed it out to Ms Guerrero.

'It's hard for them to concentrate for so long,' she said, 'so the program rewards them with a game.'

On the wall hung reminders of the behaviour expected in the Lab. FUERZA meant 'Facing forward, Undivided attention, Eyes tracking the speaker, Respectful responding, Zealous participant, All four on the floor'. LAZER was 'Line order, Arms at your sides, Zipped lips, Eyes forward, Ready to walk'. There were also motivational quotes.

> Unless someone like you cares a whole awful lot, nothing is going to get better. It's not.
>
> Dr Seuss[12]

> Whether you think you can or you can't you're right.
>
> Henry Ford[13]

It was a positive, conscientious working environment – an office for kids.

Once Martha had brought the shuttle in to land, the computer presented her with a further problem. It was winter. She had stockpiled ten snowballs and now threw eight at her friend. How many did she have remaining? The screen presented four ways of visualising the problem. First a number chart from 1 to 10. Martha quickly clicked on each number, 1, 2, 3, 4, 5, 6, 7, 8, then stopped. For each click she received a green tick. Then a box with two rows, each of five snowballs. Again, she clicked all of those on the top row, then three on the bottom row, receiving another eight green ticks. She was working smoothly. Third, the problem was written as $10 - 8 = [\ ]$. She typed a 2. Green tick again. And lastly a written problem: if I take eight from ten, what do I have left? She keyed in the answer, t-w-o.

In the past a teacher would have set and administered these problems. The kids would all have answered the same ones, then swapped papers before painstakingly grading each other's work. The genius of this system was that each individual child was carrying out a set of drills that were specially tailored to their learning needs. If they had a weakness in multiplication, the software would learn of it through analysing their data, and then ensure they practised those multiplication problems in a range of different ways. If the kid was getting everything right, the software would increase the complexity of the problems. If a hint was required, or a little encouragement, it would be delivered by an onscreen avatar. No teacher was needed. Nor another kid to mark their efforts. They'd spend between 70 and 90 minutes in the Lab each day. That was a lot of problems they were getting through.

Back in Preston's office later I considered the slogan emblazoned on the purple wall of the conference room.

If a child can't learn the way we teach, maybe we should teach the way they learn.

'For us it is a time and mastery question,' he said. 'It's multiple hits. We'll teach you in class. You'll get direct instruction. Then we'll put you in a levelled group. You might have independent time with it too. Then you're going to be at the Lab, and it's going to be at your level, so you're going to get it again. You might get pulled out for tutoring in the Lab. So if you're a child who is low, you might get the same content in a different format six different times, in six different ways in a day. We don't have an analytic system that is robust enough yet, but somehow we've got to figure out what's the best modality and the most efficient. Once we can figure that out with kids, the optimisation is astonishing.'

This optimisation was part of a trend towards personalisation in education, which Preston was excited about. It was clearly working here. Kids did well at Rocketship Schools. Really well. They performed in the 90th percentile for maths for their socio-economic group in the city and in the 85th percentile in language. The tech also saved a ton of teacher time. There had been four classes in the Lab when I visited, numbering about a hundred kids. That meant six hours of expert teacher time saved in one session. The two adult supervisors keeping an eye on things were young assistants in the early stages of teacher training.

Talk of optimisation and efficiency was a little disquieting though. Was Rocketship too enthusiastic in embracing the relentless paradigm of the machine? Rocketeers were still kids, not office workers. Writing many years after the Deep Blue match, Gary Kasparov had noted that although the AI crowd was pleased with the result, they were disappointed with the way it was achieved. 'Instead of a computer that thought and played chess like a human, with human creativity and intuition, they got one that played like a machine, systematically evaluating 200 million possible moves on the chess board per second and winning with brute number-crunching force.'[14] The risk today was that advances in artificial intelligence reduced our own intellectual faculties to something more machine-like. The thin edge of the machine wedge could be, well, a little inhuman.

Preston conceded that, at the start, Rocketship were the brute number crunchers, the 'big purple gorilla'. With a software engineer in charge and kids in classrooms, it wasn't always clear whether they were running schools or building a tech empire. But today they were much more focused on the human. While their first priority was personalised learning or 'using and adapting the latest gadgets and software to help students learn more', they otherwise focused on talent development and parent power. 'We were Uber, and now we want to be Lyft,' he said, referencing the ride-sharing app's new driver-friendly competitor. On the tech side, the biggest lesson was one of marginal gains: a slightly better version of a program, more reliable data by which to make decisions, a little more time saved.

Could a system become so efficient that one day all learning could be delivered only in supervised Labs? Preston didn't think so.

'All this bullshit about "We won't have schools and kids are gonna learn from home", I just don't buy it. And I actually think it's bad.'

He looked up with a grin.

'Our kids need to learn how to socially interact. How to smack someone across the face and apologise. And the kid tells 'em, "That hurt! And I feel sad because you smacked me!" We have social norms that kids need to learn.'

I agreed. In the Lab itself, kids were developing cognitively, but only in a fairly narrow sense. Throughout Rocketship Fuerza Prep, I also saw incredible teachers doing great work with students. This was still the core of their learning. Which was lucky. I couldn't help but feel that the things kids were learning in the Lab were precisely the type of routine cognitive skills that set them up for the most easily automatable jobs like #702 Telemarketer, rather than non-routine occupations like #1 Recreational Therapist, as outlined in the Oxford Martin report into the automation of 702 human occupations. The types of skills you could drill on computers appeared to be precisely the ones that could be easily automated by those very machines. Was it a necessary step in the children's development or an opening blunder that might one day end in a checkmate for Rocketeers and others like them?

Perhaps the Robot Teachers aren't Coming After All

The bullshit that Preston was talking about had no single cheer-leader. Rather it had developed – as things were doing in the world of machine intelligence – a mind of its own.

Much of its momentum had stemmed from a prize-winning TED talk given by Sugata Mitra about the Hole in the Wall.[15] Years earlier he had been working in New Delhi at the edge of a vast slum. He'd wondered why it was that rich kids were always considered gifted – especially when it came to computers – and poor kids not. He decided to run an experiment and installed a single internet-connected computer terminal – theft-proof, monsoon-proof and probably adult-proof – in the boundary wall of the slum. As Mitra turned on the computer he was surrounded by groups of curious kids wanting to know what he was up to. To avoid influencing his experiment, he simply shrugged and went away. When he returned, he saw something remarkable had happened. 'About eight hours later, we found them browsing and teaching each other how to browse,' he recounted, 'so I said, "Well that's impossible, because – How is it possible? They don't know anything."'[16] The conclusion that he came to was startling: that, supported by the right technology, kids could teach themselves.

Out of this kernel bloomed Mitra's idea of the Self-Organising Learning Environment, or SOLE. It was a simple formula. You posed learners a big inquiry question – something that would lead them on a journey of discovery – sprinkled in a little encouragement – someone to stand behind them, and whenever they did anything, to say, 'Well, wow, I mean, how did you do that? What's the next page? Gosh, when I was your age, I could have never done that', which he termed 'the grandmother method' – then stood back and waited for the learning to happen.

Mitra was awarded that year's $1,000,000 TED Prize for his far-sighted talk. He used his acceptance speech to proclaim a new vision for school comprising computer terminals, grannies (real-life ones from other countries with time on their hands, who would beam in to support kids at the computers via Skype) and an online learning infrastructure. It would be everywhere and nowhere, accessible to any child with a device and an internet connection. The message was pretty clear: we could say goodbye to schools as we knew them and hello to the 'School in the Cloud'.

The talk contained important insights: digital technologies now allowed access to *all* online content; methodologies could be rapidly scaled up; teachers could be beamed across the world. Eagerly on his heels came the Khan Academy, a huge online library of maths tutorials originally created by a Microsoft employee, Salman Khan, to tutor his cousin in another state, now reaching millions of learners worldwide. Both projects played on an important Silicon Valley mythology: that technology has a purpose, and that this purpose is to solve the world's problems. *Welcome to our healthy, juicy, techy utopia!* Though some were sceptical about taking these projects as the sole saviours of the future of education, others were predicting a new dawn in learning. Give a kid a laptop and they'd teach themselves anything.

The Los Angeles United School District (LAUSD) decided that it would do just that. In 2013 it announced that every student in the city would receive an iPad preloaded with Pearson software. It would be one of the most ambitious roll-outs of classroom technology seen in the USA, coming in at a cool $1.3bn.

Things did not go well.

Truckloads of iPads were delivered to the pilot schools. Many sat unused in their black flight cases, with teachers ill-prepared to use them in classes. Enterprising students found hacks around the in-built learning-only lock on the devices. Ultimately, the director of the Instructional Technology Initiative was forced to release a memo to terminate the contract, stating that only 5 per cent of kids had had consistent access to the Pearson software that drove the app. Worse, emails between Apple–Pearson and the LAUSD Superintendent were uncovered showing his excitement at working with the companies – a full year before the impartial tender process was even launched.[17] The LA iPad fiasco called to mind an earlier case where every child in Thailand was given a tablet to aid their learning. The result? Thanks to a lack of teacher training, test scores got worse overall.[18]

On top of this, a little digging by experts revealed that Sugata Mitra's Hole in the Wall wasn't *exactly* as it had seemed. Whereas he had declared that 'schools are obsolete', it turned out that in fact the computer terminals which had been installed as part of the programme roll-out were typically situated at school buildings in the slums. Mitra himself argued that the Hole in the Wall terminals worked much better as part of a programme of learning administered by good teachers.

That teachers – rather than their tools – might be the deciding factor in learning seemed often to be overlooked in the hunt for technological innovation. Rapidly accelerating connectivity and computing power had hailed a pantheon of new learning gods, from flipped classrooms to adaptive environments, blended learning to personalisation. And yet the evidence wasn't stacking up. A study by the Organisation for Economic Co-operation and Development of tens of thousands of kids in more than 40 countries revealed that the more time kids spent on computers, the worse they did on certain tests.[18] Governments mistakenly expected devices to pump up productivity on their own, forgetting about teachers. Those nations that invested heavily in technology had seen 'no discernible improvement' in test scores overall. The report concluded that 'adding twenty-first-century technologies to twentieth-century learning practices will just dilute the effectiveness of teaching'.

Robots 1 Humans 1.

But the point wasn't to dismiss technology's potential in improving learning. The findings of the OECD report, Thailand study and Hole in the Wall experiment were clear. Computers *could* transform learning – but only in the hands of expert practitioners. The report suggested that, contrary to some of the more radically held beliefs, teachers mattered more than ever. But what were their practices to look like? I decided to travel down the coast to Los Angeles. The head of the iPad roll-out was the highly respected Bernadette Lucas. She'd previously been principal at Melrose Elementary School in Central LA. It was the dazzling way that her teachers had so seamlessly incorporated iPads into student learning that had inspired the district to invest their $1.3bn in hardware in the first place.

One teacher in particular was rumoured to be an iPad Jedi. He was called Mr Willis.

On the Principle of Hybridity

After pledging allegiance to the flag, the kindergarteners of Melrose Elementary in Central Los Angeles recapped some book basics – title, author and, *for a bonus point!*, dedication – and toddled to their seats. The open doors let in the warm light of an October morning.

The kids arranged themselves in groups of six. Crayons, sugar paper, pencils and rulers were laid on each table along with the students' iPads. They turned them on, gleefully showing off the selfies they had set as background images. Mr Willis adjusted the focus on the projector and asked who'd like to show their movie, as if *who wouldn't?* In his checked shirt and khaki trousers, he was a master at capturing the kids' attention. A number of hands shot up. He chose three to start: Nathan, Jade and Eduardo.

Nathan connected to the class laptop via Bluetooth and flipped open iMovie, at home with the device. He tapped play and then covered his eyes with his small hands. Even for post-millennials – kids of the so-called iGen – showing work took some getting used to.

The camera panned across the front page of *The Red Hen* and a voiceover began.

'*This*,' said his lisping voice, 'is the title.' A second shot appeared and he peeked between his fingers. 'And *this* is the *authorsname.*' The film continued for three more scenes. By the end Nathan was sitting up, entranced. His classmates were agog. That was *woah*.

He was only five years old, could not yet read or write, and had already made his first movie.

Over recess, Mr Willis excitedly explained how the youngsters in his class use their devices. 'Oh it's wonderful,' he told me. He set the kids a project of making the best possible paper aeroplanes. Then he asked the class how they would learn to do that. The first thing that they came up with was to ask their family or friends. So they did, but they found out that their family and friends didn't know that much about paper aeroplanes. Next they decided to try looking in books, but couldn't really get hold of any with information on planes. And in the ones that they could, the instructions were hard to follow.

Finally, they wondered if they could find someone their own age who could teach them, and decided they would use their iPads and Google to search for them. They found a vlog of a seven-year-old girl in Florida who loved making paper planes. 'That was the one that worked,' said Mr Willis. It was delivered in a way they could understand, and wasn't too complicated to make. The point here was that the kids had started with a purpose, a sense of their own *why*. The technology had been a secondary consideration, though ultimately they had found it to be the right tool for the job.

Mr Willis gave Nathan some feedback. 'Nathan,' he said, 'I loved the way that you framed the title in the long tracking shot, and I really like the way that you used a clear voice. Next time, I wonder if you could play the presentation so that it shows in full-screen.'

Nathan nodded.

When Eduardo's turn came he hit play on his iMovie and the camera panned long and slow across the title – the Ken Burns effect, said Mr Willis, after the great American documentary-maker – but there was no sound.

'Shall we fix that up, Eduardo? Now, where is the button to record the voice?'

Mr Willis bustled over and helped Eduardo find the button. Sitting there, with the rest of the class watching on, he rerecorded the voiceover. The cycle of failure (no voiceover), feedback (needs voiceover) and adjustment (add voiceover successfully) lasted no longer than two minutes. Moreover, it had been instructive for the whole class, as it played out on the projected screen. It's rare in our work to see learning made visible to the whole classroom full of kids – actual tangible improvements to a child's understanding – but here it was, 20 minutes into my first visit. The technology was transparent and enabling. But so was the teaching.

Mr Willis was a veteran, one of those teachers whose work – as much as the promise of the gleaming technological hardware – inspires whole school districts to buy tablets for all their students. But it was not so much what the machine could do, impressive though it might be. It was who was using it and what he was doing. The jargon for this melding of man and machine is *hybridity*.

If there had been something eerie about Rocketship's Learning Lab – though it was doing a great job – here the iPads felt as natural as the sugar paper and crayons that littered the desks. Rather than an office, it looked like a design studio or tech start-up. Kids batted ideas around, used whatever media made sense, and were open and thoughtful in their critiques of one another's work. No matter that they were five years old. The technology felt secondary to the learning aim, and its use was transparent.

The current head teacher was a Mr Needleman. He'd come to the school from a role at the district office, taking over when Ms Lucas had left to oversee the iPad roll-out. Fundamentally, he felt, there were 'right and wrong ways of using technology. Schools shouldn't ask, "How do I use this tool?" They need to ask, "What am I trying to achieve, can a tool help?"' You couldn't just drop off iPads at schools and expect learning to improve, and you shouldn't obsess about the technology. Instead you should obsess about the teacher, and achieving the learning aim.

'People are crazy here about flipping the classroom,' he continued, talking about a new trend for teachers to video their lesson content for kids to watch for homework, freeing up class time for discussion and problem solving, 'but they are always focused on the wrong side of that – the flipped part – like how do you make a movie. People think, "Oh let me just record my lessons, kids can watch them at home." The part you don't talk about in technology workshops is what do you do now in the lesson. You've lectured them, but what do you do in class that's different? That's the part that makes the difference.'

Nonetheless, I was haunted by another realisation from Mr Willis's class. The five-year-olds I'd seen understood the *iPad* as the first portal into learning. There it was on the desk all the time, designed by the biggest company in the world and put together at Foxconn, whose CEO, Terry Gou, had publicly stated that he hoped to replace his entire workforce with robots in the next two decades.[20] Pens, pencils and books seemed noble in comparison, if a little powerless. While it felt certain that computers would have a revolutionary effect on learning, this revolution carried real dangers.

Had we really got our heads around classrooms where kids had no pencils and no books? And did we know what children should usefully be learning if the robots were taking over the jobs – even those of teachers?

I wondered what Mrs Calcutt, my Year 1 teacher with her chalk and sharing copies of *Letterland*, would make of it all.

A Final Chess Lesson

The power of hybridity became clear through an intriguing evolution in human vs computer chess. By the late 2000s you could download chess apps on your smartphone that would defeat a human grandmaster with ease and the competitions appeared to be heading down a cul-de-sac. Moore's Law dictated that processing speed would continue to double every eighteen months, as the space needed to store the same power was halved. The only semi-interesting question that remained was which device would be next to defeat a grandmaster. A smartwatch? A satnav? An alarm clock? The organisers tore up the rulebook: from now on any combination of man and machine could enter a freestyle tournament.

Gary Kasparov closely observed the evolution of these lucrative contests and reported on one for the *New York Review of Books*. Several interesting combinations of strong grandmasters and multiple computers were enticed into the competition, with the results following a predictable pattern: the teams of humans and machines dominated even the strongest computers. Even Hydra – a 'chess-specific supercomputer' and Deep Blue's most powerful descendant – was no match for a strong human player using a laptop. As Kasparov put it, 'Human strategic guidance combined with the tactical acuity of a computer was overwhelming.' But there was a surprise in store. He wrote:

> The winner was revealed to be not a grandmaster with a state-of-the-art PC but a pair of amateur American chess players using three computers at the same time. Their skill at manipulating and 'coaching' their computers to look very deeply into positions effectively counteracted the superior chess understanding of their grandmaster opponents and the greater computational power of other participants. Weak human + machine + better process was superior to a strong computer alone and, more remarkably, superior to a strong human + machine + inferior process.[21]

The chess contest had been decided not by the ablest minds, but by the best combination of cognitive powers – the superior hybrid. This anecdote is recounted by futurist economists Eric Brynjolfsson and Andrew McAfee in their book *Race Against the Machine*. In it, they lay out some stark stakes, namely – as the British writer and chronicler of technology John Lanchester has put it – that *the robots are coming and they're going to eat all the jobs*.[22] But they also take an optimistic view, seeing the result as evidence that we educators still have work to do. Humans aren't about to be overwhelmed by the intelligence of machines just yet.

This is largely due to a couple of factors. The first is Moravec's paradox, which says that the things which come easiest to us – walking, tying a shoelace or recognising a face – are the hardest to replicate mechanically (hence there is currently no robot butler to clean your house and make your breakfast), while the hardest things for us – analysing enormous data sets or working out pi to 100,000 decimal points – are fantastically simple for computers.[23] One theory is that this is all down to evolution. Take catching. It's really, *really* difficult to teach a robot to catch (though becoming rapidly less so). For us, our senses honed in the millions of years of forest living of our forebears, it takes only a little practice.

The second is that while machines can currently develop an incredible level of ability in one particular skill – like playing chess – they have no general intelligence. Deep Blue would not score a single point on *Jeopardy!*. Watson could not make the most basic first move in a game of chess. They are programmed for a single task, and can excel at it to a level that is impossible for us. Meanwhile we humans are designed for multipurpose use, carrying out a wide and varied array of simultaneous operations all the time. This is small consolation though. There are plenty of people – like Ray Kurzweil at SU – who speculate that computers will achieve general intelligence within the next 20 years, thanks to a new type of 'Strong' AI, which we'll come to later in the book. Then all bets will be off.

For now, there are serious implications for how we learn. The chess lesson teaches us that human plus machine plus good process is stronger than the best machine. But what do we mean by good process? What types of machines should our kids be equipped with? And what skills must they develop to maintain a competitive advantage? It's not time to panic – yet – but we should begin obsessing over these questions right away.

The tech prophets have already begun to. Having studied the latest technology, from driverless cars to automatic translating software, Brynjolfsson and McAfee theorised three areas where – for now – we hold the advantage: 'Ideation', the ability to come up with ideas, be creative or have a sense of purpose; 'Complex Communication', the ability to talk or write, listen or read, in highly sophisticated ways; and 'Large Frame Pattern Recognition', the ability to process a huge amount of multi-sensory information simultaneously and respond appropriately to it. Creativity, complex communication and critical thinking. It sounds like a decent blueprint for schools.

The challenge that we face today is that these skills are not being mastered by the vast majority of kids. The GCSE English course that my Year 10s waded through did equip them with some level of complex communication (they could write basic essays), but they got their C grades without truly being able to come up with new ideas, use knowledge creatively or exercise skills in novel combinations. Instead, students were on a conveyor belt towards the mastery of reading, 'riting and 'rithmetic, with many unable even to do those well when they left school. Their personal computers – in their hands they had a calculator, encyclopaedia, auto-correct, video player – could spell and add much better than they'd ever be able to.

It wasn't yet clear to me quite how powerful those machines might one day become, but it was obvious that ignoring them wasn't a great idea. We didn't want generations of kids to end up like Ken Jennings or Dr Hubert Dreyfus. Instead, we had to ensure they grew up using the latest technologies as tools for achieving their own aims. But what would that mean they should be doing at school to make the most of their human talents? It was with this question in mind that I'd driven up to Mountain View that morning in October to see Brett Schilke.

How I Learned to Stop Worrying and Love AI

'Well, here we are,' said Schilke, pausing when we reached the door of the Classroom at SU. He tapped in his entry code. Nothing happened. Technology, it seemed, was not without a sense of humour. After tracking down the new combination for the digital lock, we pushed our way in. Looking around the room at the computer monitors, 3D printers, robots and drones, I thought of that paradoxical corner of toy shops stocking gadgets for teenagers – who would still be going to toy shops at that age? – that those of us into Top Trumps, Lego and football never made it to.

Leaning against a wall was a half-size model of Darth Vader, lightsaber and all.

The Classroom was in fact a sort of lab. In it, SU students and staff could play with the latest gadgets that other companies were producing. I tried on a VR headset. One day, we might imagine ourselves seated in Einstein's lectures on relativity at Princeton, 3D renderings of his thought-experiments flashing about the virtual cosmos above our heads, but for now it was a *Honey I Shrunk the Kids* rollercoaster ride through a suburban living room. Cool, as Schilke would say, but hardly a learning revolution.

Nearby on a workbench were a series of 3D printers that could create almost any object from a variety of plastics or metals, and linked to this a device that allowed users to sculpt these materials virtually by moving their hands in the air above a bank of sensors, like Tom Cruise in *Minority Report*.[24] There was also a set of updated *Robo sapiens* robots – that weren't currently working – and a tele-presence device that looked like a Segway with an iPad strapped to the handlebars.* But I was disappointed. I'd gone in hoping for a glimpse at the distant future, where knowledge would download direct to our brains, or robot teachers would take charge of groups of kids in the way they were now looking after the elderly in Japan.[25] But this stuff? I'd seen and heard it all before.

I thought about Thomas Edison and his prediction that motion pictures would revolutionise learning. It seemed that revolutions in learning felt a little slow while they were taking place. Wasn't it true that writing had taken thousands of years – and books hundreds – to become ubiquitous in our cultures?

Buoyed by years in the Valley, Schilke felt sure that this time it would be faster.

'Education is *the* hot topic right now,' he said. 'It's going to *explode* in the next couple of years.'

The richest scions of Silicon Valley were pouring considerable amounts of money and brainpower into it. Tesla's über-futurist CEO, Elon Musk, was using an experimental unschooling approach to bring up his own kids and a few select others.[26] Just up the road in San Francisco, AltSchool had been set up by a former Google

---

* These emblems of the sometimes underwhelming reality of techno-utopianism are respectively a baby-chimp-sized toy robot that was the must-have Christmas present in 2015, and a device for beaming people's faces into corporate meeting rooms that is little more than a stick on wheels with a video screen mounted at human chest height.

executive, Max Ventilla, and raised $150,000,000 in funding on the promise of putting the latest advances in personalisation software to the problem of how to meet the needs of individual kids.[27] Teachers referred to it ominously as 'the Company'.[28] There were further rumours that Facebook had a school in the pipeline, and Google already had its feted early-childhood centre, where employees could drop their kids off for world-leading care.

The whole Valley was a crucible of educational – and technological – experimentation. The billionaires of Silicon Valley were growing up, having kids and finding that the schools available to them sucked. But they had the cash – and sense of exceptionalism – needed to do something about it. The tech-utopian's problem-solving abilities and desire to solve ever bigger problems seemed to feed off one another. If not them, then who?

Schilke continued.

'We're starting to see that we're not really preparing anyone for anything. Why would you put kids in a classroom and make them stay in this one room with one person talking to them for twelve years when that will never happen to them in their entire lives? It just makes no sense.'

The first thing was to get clear about what kids *should* be spending their time on.

'There is no reason we should be teaching spelling or math. When is the last time you added something in your head? I mean, you whip out the phone and you type it in. If you don't know how to spell a word, it doesn't really matter, Google can do a pretty good job of auto-correct.'

I had to admit the machines were getting better than us at these things. But weren't these basic skills the building blocks of our higher abilities? Didn't you have to lay the foundations in order to be able to combine those abilities in interesting and unique ways that machines perhaps never could? After my journey along the West Coast, I could now picture a school without pencil and paper. Even, perhaps, a school without books. But surely that didn't mean an end to learning to spell or multiply. I realised I had to investigate this further, to journey on to find out how our own minds functioned, how they mimicked or differed from machines.

Rather than listing off a set of twenty-first-century skills that we should therefore build into the curriculum, Schilke had a more Silicon Valley definition for the schools of the future.

'I believe the purpose of school – or learning – is to align your strengths with the needs and opportunities of the world. That's it. That's what we *should* go to school for.'

It sounded a little trite. But there was truth in it. That *was* what we should go to school for, as long as those opportunities included imagination, arts and the pursuit of knowledge for its own sake. It was hard to reconcile this sentence with the other idea: that we were in a race against the machines. But then to Schilke there was no race. The inevitability of the progress of AI was so clear to him that he was thinking in a whole different sphere. The Kool Aid could have that effect.

He spoke with calm authority, as if his parents had been preachers, as well as teachers.

'Ultimately all of this,' he said, gesturing at the shelves of hardware, 'is just allowing us to be more human.'

It was a comforting conclusion, but it rang a little false. He'd also said that Facebook knew that if they ultimately want to rule the world 'they need to have millions more people there that are crazy, talented, smart, innovative big thinkers. So there's an incentive for them to build a school in their backyard that produces that.' Now he was suggesting that in the future 'we will only be here to feed and love each other. It's about helping people be people. That's what we can do. We can determine our purpose. We can share and teach and coach and mentor.'

It was all very California.

Silicon Valley's tech-prophets were driving a revolution that was changing how the world worked. And they were pushing a new vision of education. No problem more complex – none more worthy of a solution – than that of revolutionising learning. The sociologist Robert Putnam wrote that after the land ran out in the West, the American Dream turned inwards, stretching out its wagon trains to the horizons of the mind.[29] You no longer dreamed of making it through speculating on a piece of land, you dreamed of making it through school and college. The gazillionaires of Silicon Valley already had it made. They were minted. Now they were going after the minds of the future.

As I walked back out to my rental car, looking forlorn among the Teslas and Priuses in the parking lot, I had a lot to ponder. I thought Schilke's insistence on developing our humanity made sense. It also seemed right that we shouldn't bust a gut to learn skills that computers would be able to do better than us. We should be learning how to work with machines, embracing – for want of a better word – *hybridity*. But then I didn't agree that this should be available to just some kids. The sense of outsiderism in the Valley was strong, and I wondered if it was skewing their vision.

I started the engine. How many driverless cars might I speed by on the freeway?

Human plus machine plus strong process wins. Our schools ought at least to bear that in mind, for now. In our rush to make progress, it seemed that we were overly focused on upgrading our machines. What if we applied the same effort to growing human minds? At Walworth my kids hadn't got anywhere close to the frontier of their intelligence. We had to better explore those vast wilds. After all, wasn't every piece of technology the result of human imagination? Across the world, a few hardy pioneers were breaking new ground in understanding our natural intelligence. Particular progress was being made in the realm of early-childhood development, with the infant brain beginning to reveal its secrets. I'd heard that 'baby labs' were springing up at the world's leading universities. That had to be the next stop on my journey.

'There are basic skills that we still need,' Schilke had said. We did. We do. School in an era of exponential technology was about learning to command the tools that would enable us to live our version of the good life, while maintaining our intellectual advantage. In my mind, the painting above Schilke's head took on a new hue. The android biker seemed almost human. He was hitting the open road, the wind in his ersatz chrome hair.

'Ultimately you want to understand what your AI is doing,' he added. 'Or you run the risk of the world being taken over by robots.'

In Boston, an AI expert was pitting the infant human mind against the emerging machine one. I flew to pay him a visit.

CHAPTER 2

Born to Learn

Taking Baby Steps

> You were not born to live like animals,
> but to pursue virtue and possess knowledge.
>
> Dante

In the Beginning was the Word

Deb Roy and Rupal Patel pulled up to their driveway on a fine July day in 2005 with the beaming smiles and sleep-deprived glow common to all first-time parents. Pausing in the hallway of their suburban Boston home for Grandpa to snap a photo, they chattered happily above the precious newborn son now swaddled between them in a powder-blue car seat, aware only of the familiar sound of their voices. The two university professors weren't *exactly* like other mums and dads though. Roy was an AI and robotics expert at MIT, Patel an eminent speech and language specialist at nearby Northeastern. They'd been plotting. If junior's eyesight had been further along, he'd have spied two discrete black dots, each the size of a coin, blinking from the ceiling. Further dots hovered over the open-plan living area and the dining room. Through the kitchen, bathrooms, hallways and bedrooms, there were twenty-five in all. They were fourteen microphones and eleven fish-eye cameras shipped from Japan, part of a system primed to launch on their return from hospital, designed to record the newborn's every move.

These average-looking suburban parents were planning to amass the most extensive home video collection ever. They'd dubbed it *Total Recall*.

Roy and Patel had met a decade earlier as fellow explorers at the frontiers of the science of mind. Roy was six years old when he built his first robots back in Winnipeg, Manitoba in the 1970s and he'd never really stopped. With no ghost in his early machines, which were purely cosmetic – like baddies in early sci-fi movies – he'd wondered about android brains. What would it mean for his machines to think? 'I thought I could just read the literature on how kids do it, and that would give me a blueprint for building my language and learning robots,' he told *Wired* magazine.[1] Over dinner one night he boasted to Patel, then completing her PhD in human speech pathology, that he had created a robot that was learning *the way kids learn*. 'I bet,' he said, 'that if we gave it the sort of input that kids get the robot could learn from it.'

Toco, the best of his learning machines, was little more than a camera and microphone mounted on a Meccano frame and brought to life *Blue Peter*-style with ping-pong-ball eyes, a red feather quiff and crooked yellow bill. But it was smart. Using voice recognition and pattern-analysing algorithms, Roy had painstakingly taught Toco to distinguish words and concepts within the maelstrom of everyday speech, a huge breakthrough. Previously computers learned language digitally, understanding words in relation to other words. Here for the first time was a machine that understood their relationship to objects. Told to pick out the red ball among a range of physical items, Toco *could*.

Patel told Roy to prove it, and so he flew to her infant lab in Toronto to get started. Observing the mothers and babies at play, he realised he'd been teaching Toco badly. 'I hadn't structured my learning algorithm correctly,' he explained at the time. 'Every parent knows that when you're talking to an eleven-month-old, you stay on a very tight subject. If you're talking about a cup, you stick to a cup and you interact with the cup until the baby gets bored and then the cup goes away.'[2] Where his robot had previously searched through every phoneme* it had ever heard when learning, Roy tweaked its

---

* Phonemes are the distinct units of sound that make up a language. It's how you

algorithm to give extra weight to its most recent experiences and fed it audio taken directly from Patel's baby lab recordings. Suddenly Toco began to build a basic vocabulary at a rate never seen before in AI research. His dream of 'a robot that can learn by listening and seeing objects' felt closer than ever. It needed to feed on recordings though, and these were hard to find. 'Trying to get realistic samples of what a child actually learns turned out to be a bottleneck.'[3]

No one had *ever* truly studied 'in the wild' what happened to a child during those first crucial years. The weekly hour-long sessions of observation that Patel used to study mothers and infants in her lab were the norm. Some pioneering psychologists had tried studying their own children, like Michael Tomasello, a cognitive scientist, who gave up one day when his daughter turned to him after doing something praiseworthy to ask, 'Are you going to write that in your book, Daddy?' The trouble was that the presence of an observer changed things. Either that, or you had to infer a lot from a very small sample. But language didn't occur in a vacuum, however much our schools now treated it as though it did. 'I had decided,' wrote Jerome Bruner, the godfather of cognitive psychology, 'that you could only study language acquisition at home, in vivo, not in the lab, in vitro.'[4] If you were going to study the way a baby learned to talk, you'd need someone eccentric enough to rig up a house with hidden recording devices. Enter Patel and Roy.

The new parents agreed some ground rules. The recordings would be available only to their most trusted inner circle of researchers. If at any time they no longer felt comfortable with the filming, they would junk the footage. 'Oops' buttons were installed in every room which could erase recently filmed material. When privacy was required, the system could be temporarily shut down. The aim was to dial down the voyeurism, dial up the science. 'It's the difference between being recorded and recording yourself,' Roy told me when I visited him at his Boston lab. It was a leap of faith, but they agreed it

---

pronounce letters or combinations of letters when you actually say words. So p, d, b or t in pad, pat, bad or bat, rather than how you'd pronounce them when reciting the alphabet. English has 44 phonemes – and a very strange and unintuitive way of pronouncing its alphabet.

was worth it. Their experiment had the power to unlock new insight into the workings of the infant mind.

'How does this language-learning thing work for one kid?' said Roy. 'Surely it's not just words that matter but context?'

Toco was Pinocchio to Roy's Geppetto. But whereas he was wondering what real kids could teach robots, I wanted to know if those home videos might hint at how to enhance learning for the youngest humans. Science now showed how important this was.

No, Your Baby Can't Read

In 1985, too late for my own early childhood, two researchers, Betty Hart and Todd Risley, published the results of a study in which they trailed 42 Kansas City families to compare the experiences of wrong-side-of-the-tracks preschoolers with their richer most-likely-to-succeed peers. Starting when the infants were nine months old, they observed them regularly over a two-and-a-half-year period, recording and transcribing all parent and child speech during their hour-long visits. The findings were stark. The number of words a child heard by their third birthday strongly predicted academic success aged nine. The difference was barely fathomable. They estimated that at the age of four the richest kids had heard *30 million* more words than the poorest.[5]

'The problem of skill differences among children at the time of school entry is bigger, more intractable, and more important than we thought,' Hart and Risley warned. You had to intervene as early as possible. 'The longer the effort is put off, the less possible change becomes.'[6] Megaphoned out in the media – *The 30 Million Word Gap!* – their findings fuelled a word-rush. Parents flocked to buy flashcards and brain trainers. Governments funnelled funds towards formal nursery tuition and administered tests to ever younger infants to evaluate their 'school-readiness'. Later, new techniques enabled scientists to perceive a six-month language acquisition lag between the richest and poorest babies by the age of just one and a half.[7] The gap was cumulative and it started at birth. You had to fill it with words.

It seemed to be just as Roy had hypothesised with Toco, but to

me the interpretation felt a little simplistic. Yes, we had to close
the achievement gap, but surely there was more to infant learning
than words? A professor of early-childhood development at Temple
University in Pennsylvania had written that 'just as the fast food
industry fills us with empty calories, what we call the "learning industry" has convinced many among us that the memorisation of
content is all that is needed for learning success and joyful lives'.[8]
She'd also written an influential book that laid out her reservations
with the word-rush: *Einstein Never Used Flashcards: How Children
Really Learn and Why They Need to Play More and Memorize Less.*
I thought that she might have some answers.

Her Philadelphia lab was just a short journey from Boston, so
with the hope of finding out how the infant mind *really* learned, I
took the Amtrak down to meet her.

The Scientist in the Crib

'Bah doh gah bi bah boh we gah *bah*,' chanted Kathy Hirsh-Pasek,
as we sat in her kitchen in Ardmore, a leafy Philadelphia suburb,
breakfasting on potato latkes. Her sprawling home was alive with
final preparations for a fundraiser that evening. Decorators came
and went. Friends called to gossip about the state of the neighbourhood, and the nation. Hirsh-Pasek was the energy source at the
centre of a busy universe. People were her passion (when I called for
an interview, she promptly invited me to stay for two days) as well
as her expertise. The author of 11 books and 150 academic articles,
she was legendary in the field of early child development, a Distinguished Faculty Fellow who ran the Infant Language Laboratory at
Temple University, its slogan: *Where children teach adults!*

I'd heard that in the lab scientists and eager grad students put
tiny humans through their paces. We were going to drive over
later.

'Doh gah doh boh bah si dey mi gah mi lah,' she continued,
modulating her voice into a looping monotone.

The strange performance was a re-enactment of research into the
infant mind carried out by her friend and colleague Jenny Saffran,
who had read eight-month-old infants two-minute-long screeds of

the seemingly nonsensical babble that Hirsh-Pasek was parroting. It wasn't total gobbledegook though. Within the flurry of sounds, certain phonemes – in this case 'gah' – were repeated more often than others. The genius of infant brains was that they *noticed* these recurrences. Their brains fed on repetition. Saffran termed it statistical learning.[9] Others like the psychologist Alison Gopnik called human babies 'the scientists in the crib'.[10] The infant brain was the world's most powerful learning device. We were *natural born learners*.

'Vah si zi meh boh di beh li gah *deh*,' she exhaled, finally running out of puff.

'Babies pick up those probabilities,' she exclaimed, wide-eyed. 'When they hear "pretty baby", they know that the "pre" and "tty" go together, but the "ty" and "ba" don't. How do they do that?' She was in awe of the infant mind and I was starting to see why. Saffran's ingenious experiments measured baby listening times or changes in heart rate to show a number of things that eight-month-olds already knew. 'They know the mobile won't fall on them,' said Hirsh-Pasek. 'They know that if I drop this plate on the table, the plate won't go through the table. *That's amazing.* They know that if I'm sitting across from you, and you can't see the bottom part of my body, do I have one?' Hesitantly, I answered yes.

'Babies expect that too! That's *so cool*!'

It *was* remarkable. Until the 1990s we'd thought of infants as irrational, illogical and egocentric. In 1890 William James, brother of Henry and with some of his way with words, had written in his *Principles of Psychology* that 'The baby, assailed by eyes, ears, nose, skin, and entrails at once, feels it all as one great blooming, buzzing confusion', painting an image of sensory overload.

But it wasn't true.

'You look into the eyes of a baby,' said Hirsh-Pasek, lowering her voice reverentially. 'And they're already pattern-makers. It's unbelievable.'

Even in utero, unborn kids picked up sounds. One-hour-olds could distinguish their mother's voice from another person's.[11] They arrived in the world knowing which family of languages they belonged to, English say, over Japanese, and equipped with a set of abilities bequeathed by evolution, like sucking or breathing, along with a brain primed to learn through sensory stimulation and a proto-scientific method. Using every one of their senses a newborn stretched out into the world 'trying to adapt the very limited suite of skills it has to be able to get more information and then home in on it', she added.

We were born explorers ready-made for scientific inquiry. We had to understand this if we were to realise our learning potential.

'For the most part evolution has prepared us well,' said Hirsh-Pasek. We entered the world ready to 'read the perfect cues out of the environment', she explained. I thought back to Toco. He read the environment too. Well, at least what his eye cameras saw and ear microphones heard. But he and other robots could only reach out in ways they'd been programmed to, could only learn from stimuli they were instructed to pay attention to. It limited them to a small range of experiences that would shape their behaviours. There was no meaning in their methods. Babies on the other hand used scientific inquiry. And they had a further superpower. They were *social* learners.

'We're as shaped as we are shapers,' continued Hirsh-Pasek. 'We arrive ready to interact with other humans and our culture.'

The real genius of human babies was not simply that they learned from the environment – other animals could do that. Instead, our infants had a unique trick up their sleeves. They could read minds.

On the Origins of Human Cognition

The rise of the machines had occluded an important strand of research into our minds, which focused on the abilities we shared – and those we didn't – with our primate cousins. At the Max Planck Institute for Evolutionary Anthropology in Leipzig, Professor Michael Tomasello spent decades studying humans and chimpanzees (with whom we share about 99 per cent of our genetic material) from birth to 24 months in order to analyse how we learned. Our understanding of the material world appeared to differ little. What psychologists had shown for humans, Tomasello proved for chimps: at four months, both species could reach for and grasp objects; at eight months, we'd look for objects that disappeared; from a year we categorised, estimated small quantities, perceptually rotated, got what it meant to hide something; and at a year and a half grasped the spatial, temporal and causal relationships between objects. We understood snooker in other words, minus the waistcoats.

There was only one significant way in which our species diverged: our perception of the minds of others.

For 2 million years, wrote Tomasello, our whole *Homo* lineage had no more mindpower than that of other great apes. Then at an unknown juncture around 200,000 years ago – a blink of the eye evolutionarily – our species underwent a cognitive revolution. Suddenly, we made new stone tools. We shaped symbols to communicate and created new forms of expression in cave paintings or stone carvings. Cultural rituals emerged, from burying the dead to domesticating plants and animals – the roots of today's computers, writing and religions. But why did this happen? Tomasello believed it relied on the arrival of human beings' particular superpower: the understanding of other humans as *being like us* and, most crucially, of having mental lives like our own, specifically their *own intentions*.[12]

We had evolved a social sixth sense. Social and cultural transmission became possible.[13]

Language was our starting point, the possibility of two beings ascribing a shared meaning to an otherwise abstract concept or symbol. This gave rise to culture. Our hominid ancestors could suddenly interpret the environment and artefacts around them as resulting from the creative actions of another. Couldn't we see shoots of this in the behaviour of babies from day one? Infants under a year engaged in proto-conversations with carers. They babbled away, held eye contact with Mum and Dad, exchanged things, mimicked their expressions or actions. They also experimented with *tools*, sticking them in their mouths, bashing them on things.

Tomasello wrote of 'the cultural ratchet effect'. In contrast to other animals, humans were able to stand on the shoulders of the generations that came before. Our young learned 'in an environment of ever-new artefacts and social practices which, at any one time, represent something resembling the entire collective wisdom of the entire social group throughout its entire cultural history'.[14]

If all of us were to achieve our potential as learners, the question we had to answer was how we ought to shape this environment.

Our brains had specially adapted to learn. Our long period of immaturity was a risky evolutionary strategy, making us vulnerable early on to predators or sickness and delaying for many years our capacity to reproduce, but the payoff was immense. We could actively incorporate enormous amounts of the latest information from our environment and social group into our cognitive development.[15] Learning even prompted changes in our genes. Scientists have long recognised the nature-versus-nurture debate as fallacy. Genes are now known to ascribe a direction of travel and set limits on our growth (a baby chimp can't grow into a human kid, however much it wants to), but they are expressed only in interaction with the environment.

This places the utmost importance on learning in our earliest years. A huge amount of our brain development takes place in the first three years of our lives. From birth to our thirty-sixth month, our grey matter grows in mass from 0.3 to 1.3 kilos. In that time, some of our most important cognitive architecture is laid down, especially for speech. In famous cases of children left alone or brought up in the wild away from human language for the first three years, the kids have often been unable ever to learn to talk, having moved past the point where it was physiologically possible.[16] In those first years, the brain grows in relation to the environment, forming itself in interaction with near-infinite sensory experience.

As Hart and Risley showed in their study of the word gap, that experience had a huge effect on who that person became.

We'd *evolved* to be a species of teachers and learners. Our mind-reading ability arrived with what Tomasello termed the 'nine-month revolution', a moment in their development at which babies began to check the attention of others by holding or pointing at objects. At a year they could follow another's attention, gazing at, touching or listening to the same thing. At 15 months they could direct it. *Listen to that! Look over there!* Shared attention was the starting point of conscious human learning. It was why infants didn't learn to talk from video, audio or overhearing parental conversations. We hadn't evolved to. It was why it mattered whether we talked with our kids. It was also why we couldn't learn from robots. Yet.

The implication for our schools sounded a lot like common sense and a little bit New Age. Each generation ought to ensure that the next was steeped in their earliest years in the tools, symbols and social practices of the current culture. I'd driven out one day to a former steel town in the British Midlands to visit a place that was trying to do just that. What kind of learning environment might best cultivate our natural born abilities?

At the Human Nature Reserve

The Beach was cold and overcast, but that hadn't deterred the crowds. Two small boys splashed with sandcastle moulds and pipes at an ever-running tap by the bamboo bush. 'Don't get me wet!' they squeaked with delight. A teacher bent down to comfort a toddler whose puffa, wellies and *Be Fast or Come Last* T-shirt advertised his early passion for cars. Four small girls were deep in a serious conversation while absent-mindedly digging sand into colourful buckets. On the wooden walkway that enclosed the terrain on four sides, two boys jogged back and forth on an imagined errand. The setting was the Pen Green Early Childhood Centre, where kids were in their natural habitat.

A three-year-old Batman knelt with the Joker beneath the boughs of a tree, plotting. Between them crouched Superman, whose face was painted to look like Death.

Alongside the bamboo thicket, the two engineers gleefully dug out a channel in the sand, coaxing the water towards a home-made oasis.

'It's a waterfall!' they cheered.

Car Kid headed over to investigate. They were three pint-sized scientists investigating forces.

'Come back *now*!' yelled Superman-Slash-Death. It was his birthday and like his two partners in crime he remained firmly in character. He roared suddenly. Batman and the Joker split and he followed close on their heels.

'Come back *now*!' he repeated, running behind at never-quite-catching-up-speed and growling as they lapped the Beach, jogging round the wooden terrace, out through the oasis, behind the bamboo and in and out of the cabin of a shipwrecked vessel.

A teacher crouched to hug a blonde-haired boy. He was in tears, pointing at a pair of troublemakers who were throwing sand. She led him over to resolve the situation together. What had happened? What if Blonde Hair joined their game?

'Come back *now*!' hollered Superman-Slash-Death, as the four girls continued to dig. What was going on in these kids' heads?

Earlier, I'd spoken to Angela Prodger, director of Pen Green. She'd just taken over from the legendary Margy Whalley, who'd set the centre up 35 years earlier after running programmes in Brazil and Papua New Guinea. Established in the Northamptonshire town of Corby, a couple of hours north of London, Pen Green had a global reputation for excellence in early child development and family support, a prototype that had inspired successive government interventions like Sure Start and Early Excellence.* In the 1980s, Corby had been among the UK's poorest and least educated places, its population of Scottish migrant workers unmoored by the closure of the steelworks for which they'd moved south, which saw 11,000 people made redundant. The centre was intended as a lifeline for the next generation. Today it serves 1,400 of the UK's least well-off households.

'We see ourselves as a community hub,' said Prodger. Sure enough, as the day began dads in hi-vis jackets and work boots trooped in alongside mums with prams, toddlers skulking about their knees until the double doors opened and they streamed off to play. The centre offered a range of services. There were three nurseries for two- to four-year-olds, two zones for nine-month- to two-year-olds, crèche facilities for families experiencing challenges and support services for the most vulnerable parents. Round every corner was a volunteer or practitioner – all female – with a happy story of sending their own tot to the centre, then somehow finding themselves still

---

* Sure Start centres were a UK government initiative launched at the turn of the millennium to give free access for the kids of all poor families to a centre in their local community at which they and their parents would receive the support and resources needed to help them overcome the developmental gaps they experienced in relation to their more affluent peers. A decade later, 3,500 such centres existed to give disadvantaged kids a leg-up early in life, though many have now closed or reduced their offer thanks to funding cuts following the financial crisis of 2008.

here after 20 years, the kids long since grown up. Pen Green gave them a chance to earn degrees at the attached research centre, training up parents as they were bringing up their kids.

I asked about the language learning of the kids. We knew words mattered, but I'd not heard much talk.

'If we're not addressing personal, social, emotional development first, you're not ready to learn,' said Prodger patiently. You had to understand that before kids could acquire the tools of speech and language, you had to ensure they felt a sense of 'being and belonging'. Too frequently, she thought, our approaches to learning skipped these steps. It sounded to me like a nice-to-have, but research had begun to show otherwise.

In the 1950s a British psychoanalyst, John Bowlby, proposed a theory of 'attachment' based on a series of experiments carried out on baby rhesus monkeys by psychologist Harry Harlow. Harlow placed infant primates inside a cage with two mother monkey puppets – one was covered in fur, the other just a wire skeleton – both of which were equipped with a fake nipple from which the infant could feed on milk. The experiments revealed that regardless of which 'mother' did the feeding, the infant primates would spend their time attached to the furry simulacrum rather than the wiry one, suggesting that the existing 'behaviourist' view (that we were attached to our mothers because they were our source of food) was wide of the mark.[17] Instead, closeness to another – Harlow talked of 'love' – was fundamental if we were to thrive as mammals. Bowlby hypothesised that unable to regulate their own feelings in infancy, infants were prone to upset when they were hungry, sad or lonely. A carer was needed to help them 'co-regulate' their feelings, which over time would teach the child to *self*-regulate, provided their early experiences helped them to do so.[18]

More recent science seemed to confirm this Attachment Theory. One study asked preteen girls to solve brain-teasers in front of a live audience. Following the stress-inducing challenge they were then allowed either to meet their mum in person, speak to her on the phone, receive a text from her (presumably all in caps, if my mum is anything to go by) or have no contact at all. Researchers then measured the levels of cortisol – a stress hormone – in the saliva of each group. Girls who saw their mother in person experienced the biggest stress reduction, thanks to their brains boosting the production of oxytocin after they saw their mums.[19] It's also been shown that this effect may be expressed epigenetically, meaning it could permanently alter a person's DNA.[20] Experiments by Professor Michael Meaney demonstrated that rat pups licked and cared for by an attentive mother rat in the first week of their lives showed greater confidence in approaching new tasks and a better ability to regulate stress throughout their entire lifetimes.[21]

Unless alleviated through love from a parental figure, negative experiences could become permanently structured into your brain. The implications for kids growing up in poverty-stricken or traumatic environments were significant. This was one of the reasons why Pen Green took care to put the being and belonging of its kids first. It also showed the behaviour displayed by the kids at Walworth Academy in a new light. Where I'd missed the signs of kids responding to the stress of the environment in which they were growing up, retaliating with detentions, at Pen Green they worked closely with carers to ensure kids built strong nurturing relationships that would help them thrive in the nursery and ultimately at school. I'd always believed kids *wanted* to wreak chaos. It had never occurred to me that they might simply have been conditioned by their environment to act in a certain way.

'Behaviour is always just a sign of children trying to tell you something,' said Prodger.

As we toured the building through the Couthy, the Nest, the Den and the Snug, Prodger explained further that the skill of the practitioners at Pen Green was in learning to attend to what was going on in the minds of the kids in their groups, and interpret it as evidence of what the youngsters were signalling, even before they were able to verbalise it themselves. Kids were constantly communicating with us. We just had to learn to understand.

'It's about looking,' Angela explained. 'What are the children trying to explore? What are they trying to find out?'

Recently, Prodger had noticed a group of boys running around playing goodies and baddies. In another environment they might be sent outside and ignored – 'They're told, boys, you're being a bit noisy, if you want to do that go and let off steam outside' – but she saw something different. Children engaging in play, acting out a story. Wondering how to use this to develop their literacy skills, the next day she had Playmobil figures ready for them. Their tiny, fiddly arms were ideal for developing kids' fine motor skills, which would in turn set them up to use pencils in the future. She asked if they could give her a story about the figures and from there led them into storybooks. 'I wonder if you could draw me a baddie?' This was the difference between forcing all four-year-olds to sit down together to work on the same task – which the US and UK governments were pushing hard for – and following the evidence, which said to let them play.

'Children are preprogrammed to learn,' Margy Whalley had told me a few months earlier when we'd met in her office. Kids don't *flit*, went one of the centre's favourite sayings, they *fit*. First, as part of a family and community and then into behavioural patterns called schemas. They'd first been noticed by Jean Piaget, then popularised by cognitive scientists today. The schemas related to urges that seemed to occur naturally in small children. An 'enveloping' schema meant kids liked to cover themselves in fabric or paint and enjoyed dressing up, while a 'trajectory' schema was all to do with running back and forth, jumping or throwing. Others included 'rotation', 'orientation', 'positioning', 'connection', 'transporting' or 'transformation'. If you've ever seen toddlers lining up toy cars or teddies, that was 'positioning'.

Developing these urges was part of the process of preparing infants for learning later in life, fuelling things like imagination, narrative ability, spatial awareness and number sense which would be the foundation on which creativity, language, maths and science would be built. If you started too early with flashcards, you supposedly lost this developmental stage.

'It's about being free,' Prodger had said. 'It's about risktaking.' They took the kids out to the forest a few days a week, lit fires, let them experiment with scissors, ride BMX bikes. They had the run of the whole of Pen Green, like contestants on *The Crystal Maze*. If they wanted to be outside, they went outside. If they fancied returning to the Couthy, where the youngest infants rolled around, that's where they headed. The environment dictated the learning. The adults aimed only to connect, to share attention with the kids. UCLA Professor Dan Siegel had called this 'mindsight'.[22] Reading and writing could wait. Nurseries ought to be as social as possible and follow kids' lead in their play. Before kids could get on with learning, we had to ensure they belonged.

Over in the Snug, the teacher, Sara, knelt down to give Batman a hug. Seeing an opportunity to develop their future literacy, she'd laid out a table of action figures for them to play with. She wanted to build on the schema in which they were characters in their own story and channel it into social, imaginative and linguistic play. Could they create a superhero story that they acted out together with the action figures? A future step would be to tell the story to a teacher, so that she might write it down. A couple of weeks earlier, Superman-Slash-Death had been on a visit to see a steam train. Today he'd sat down and told an assistant the story. I saw him running off to meet his dad, who'd come in to get him in his boiler suit and work boots. He beamed with pride. He didn't know how to read or write, but in his hand he clutched his first book.

'It takes a village to raise a child,' Margy had said. The old chestnut had a new shine to it. For the youngest kids, I thought, the village probably ought to look something like Pen Green. They were happy here, learning to belong and laying down foundations for their future success through play. And yet I wondered if we couldn't do still more to accelerate early learning. The implication of Deb Roy's experiment was that every moment counted. Could we afford to leave so much to chance?

Rethinking the Learning Curve

In contrast to most maths formulae, which live in the heads of geniuses or on blackboards at universities, the Heckman Equation has its own website and mailing list, though sadly no Instagram feed, in case you're wondering what it's been up to this week. Developed at the turn of the millennium by a Nobel Laureate economist, James Heckman, it is usually illustrated by a simple graph. On the x axis is a human lifespan from birth to death, on the y axis – slightly depressingly, but he *is* an economist – the rate of return on investment. There are other ways to weigh a life of course, but let's imagine we live in the era of late capitalism. The question Heckman asks is: what financial return does society get for each dollar invested in a child from zero to eighteen?

The steep resulting curve would be a lot of fun to ride your bike down. From age zero to age three you get *a huge* return on your investment, from four to 18 *some* return and from 19 onwards *a little*.

The equation was drawn from analysing historic early-childhood interventions from the US, among which was the Perry Preschool Program. Launched in Ypsilanti, Michigan, in the early 1960s, it saw disadvantaged three-year-olds spend two years enrolled in preschool classes that encouraged active learning, like problem solving or making choices for two and a half hours each school day. As the tots played, adults gently coaxed them to use a plan–do–review approach: to come up with their own ideas, act them out then reflect on what happened. Teachers supplemented this with a weekly home visit. It was an epic study, an early *Seven Up* or *Child of Our Time*,[23] with follow-ups carried out on Perry kids at 15, 19, 27 and 40. The numbers were inarguable. In today's money the $16,500 investment in two years of high-quality preschool generated benefits in increased lifetime earnings, reduced crime and lower welfare spending of $145,000 dollars.[24]

It seems a no-brainer. Yet Heckman's curve turns all we believe about learning on its head. As a teacher I'd known that you really showed the right stuff at A level. Facilitating talk among young adults was Socratic. What *was* it to be human on a crowded bendy bus at 6.45 a.m.? Secondary school came next. Prompting, nagging and motivating those half-remembered teenage faces towards life-defining GCSEs was true work at the chalk-face. After that was primary school. We dreamed of happy children gambolling into riotously decorated rooms, politely doing as they should and gathering excitedly on the mat for story-time. At worst you were occasionally called 'mummy'. Finally there was nursery. We knew it as glorified childcare, run by mums with time on their hands. The curve I'd known was the *opposite* of Heckman's.

'The accident of birth is the greatest source of inequality in the

US,' he wrote.[25] It was equally true in the UK. If we wanted to bring about a revolution in the way we learned, it wasn't enough to transform schools. We'd have to start much earlier than that, flipping the curve. It was with this thought in mind that I'd driven with Kathy Hirsh-Pasek to her Lab.

Where Children Teach Adults!

The Temple Infant and Child Lab was housed in a historic nineteenth-century building, or convincing replica of one, surrounded by green fields. Hirsh-Pasek and I pulled in at 10 a.m., having dropped off fresh lemonade to her friend on the way. The Lab was the result of a stellar career that had taken her from early days at Oxford, through the tutelage of Jerome 'Jerry' Bruner ('I was literally infatuated with the man,' she told me, 'I thought he was the best teacher I had ever seen'), to the writing of numerous books and papers and the winning of a whole suite of the American Psychological Association's most prestigious awards.

It was also an extension of her personality, built around piercing insight, beaming smiles and no standing on ceremony.

A triceratops, diplodocus, and stegosaurus met us as we entered the waiting area, strewn with toys for the littlest people. In front of the dinosaurs were a miniature phone box, kitchen counter, multi-coloured truck and panoply of stuffed animals ranged on top of the knee-high bookcase, home to *Curious George*, *Dr Seuss* and *The Hungry Caterpillar*. It wasn't lab-like, but homely and a little ramshackle. A cardboard box marked 'Oriental Trading' sat on top of a cupboard decorated with a cartoon flamingo and promising *fun products for every occasion*. I glimpsed signs of research through open doorways into small rooms divided with black curtains, where laptops connected to camcorders awaited their next test subjects.

On the first floor, 19 fresh-faced, casually dressed post-docs sat waiting for me with chips, dips and a PowerPoint. Dr Rebecca Alper was leading research around 'enhancing the communication foundation'. She hit 'play' on a video.

An attentive mum in a red sweater lay on her side on a carpet opposite her infant son. I recognised the zany town map rug in the video from the Lab's downstairs play area that I'd just passed through. Between them was a bowl of cereal.

Mum picked up a Cheerio and popped it in her mouth as her son watched.

'One for Mommy,' she said, gazing right into his eyes. Sitting up in a white playsuit and surrounded by debris, he looked delighted.

He reached into the bowl for a Cheerio and popped it in his mouth, then repeated the move.

'Can I have some cereal?' asked Mum.

Junior put another Cheerio in his own mouth.

'One for you, one for Mommy,' she said, picking up some cereal pieces before feeding one to him and then one to herself.

'Can you feed Mommy now?'

He watched her thoughtfully as he crunched down his mouthful then carefully pushed his hand into the dry cereal, pulling out one of the wheat circles and feeding it to Mum.

'Thank you!' she encouraged him, continuing to hold his gaze as they both crunched away.

This, explained Alper, was a conversational duet. She'd prepared the video with another post-doc at the Lab, Dr Rufan Luo. They were working with the Maternity Care Coalition, who'd long focused on ensuring the poor kids across the US got a better start in life, by working with mothers to ensure that their daughters and sons were healthy and securely attached. Alper and Luo were now training up the home visitors of the Coalition to help parents improve the quality of their speaking interactions with their infant kids. The design of the duet was rooted in babies' in-built superpower. Though infants had a scientific method, they needed elders to act as guides in shaping the focus of their investigations towards the richest sources of learning. That was the point of teachers. They increased the likelihood of meaningful discovery, could help you practise the things that mattered. The quantity of words mattered, but so too did the quality of interactions. The shared focus on the bowl of Cheerios was an example of shared attention.

The conversational duet was one of a number of experiments at the Lab, all aimed at closing developmental gaps between rich and poor kids. Others covered topics like language development, playful learning and spatial awareness, and all used technology.

I had thought the youngest kids couldn't learn from screens.

'I came up with this thing, "prompts and partners",' Hirsh-Pasek told me. 'What the machine really can't do now is be a partner. It isn't social. It's interactive without being adaptive.'

You couldn't talk to a robot, or feel cared for by one. But grown-ups could learn from them, and they could give timely reminders. The whole app industry had recently gone nuts over a paper that Kathy had written in which she proposed a simple set of rules: active over passive, 'and it can't be stupid active, like swiping something'; engaging not distracting, i.e. 'does it keep you on focus, or take you somewhere else?'; meaningful, or 'are you teaching people stuff that's never going to connect with anything that anyone is going to give a darn about?'; and socially interactive, 'is there another person there?' If it fulfilled those rules, had a learning goal and wasn't too easy – otherwise, why bother? – it qualified for the 'Hirsh-Pasek Seal'. A system like that could be pretty lucrative for her, I thought.

Failing to meet her criteria, some developers had complained that the kids really enjoyed their apps.

'Yeah, and they like dessert too,' she said, 'but you're still not willing to give them a whole *meal* of cake and cookies.'

Hirsh-Pasek's mission was to change the way we thought about learning, especially for the poorest kids. In her book she quoted Stanford University's Professor Linda Darling-Hammond. 'We want to prepare our children for the twenty-first-century world they are entering, not for an endless series of multiple-choice tests that increasingly deflect us from our mission to teach them well.' She was especially worried that we were pushing this onto the most vulnerable kids. 'We had this vision that it was so important to get the basics into poor kids. It was so important that we should drop recess. When you know being physical helps kids learn, helps build better brains. And that we should just do reading and maths, and that we should cut out the arts, we should cut out all this superflu-ous stuff like social studies.'

It weighed heavily on her. Policymakers and laymen had twisted the science to fit their own ends. No scientist thought that flashcards worked. No scientist believed that you should start learning to read and write at an ever younger age. It was a fantasy of governments.

More recent research has added depth to the language lessons of Hart and Risley's Kansas Study. In 2003, the psychologist Patricia Kuhl experimented with teaching American infants Mandarin. Split into three groups (video, audio and flesh-and-blood teacher) only those with a human tutor learned anything at all.[26] In 2010 a study of the wildly popular Baby Einstein vocabulary-building DVDs (*Time* had called them 'Crack for Babies')[27] revealed that infants who watched them 'showed no greater understanding of words from the program than kids who never saw it'.[28] Nor did babies learn words by eavesdropping on parental conversations or listening to *In Our Time* on Radio 4, however soothing the mellifluous tones of Melvyn Bragg. More than words, it took a human being for a baby to learn language. They couldn't learn from screens.

Schools were still ignoring this secret source of infant learning. Erica Christakis at Yale had charted the slow descent in preschool learning from a multidimensional ideas-based approach to a two-dimensional naming-and-labelling curriculum.[29] Hirsh-Pasek's friend Daphna Bassok had done some research at the University of Virginia asking if kindergarten really was the new First Grade. The answer was yes.[30] Arts were down 16 per cent, testing up 29 per cent. The expectation that kindergarteners could read was now commonplace. Yet it was counter to all the evidence. A Cambridge study comparing groups of children who started formal literacy lessons at five and seven found that starting two years earlier made no difference at all to a child's reading ability aged 11, 'but the children who started at five developed less-positive attitudes to reading, and showed poorer text comprehension than those who started later'.[31] An analysis of international school data showed no relationship between when you started school, whether aged four or seven, and your eventual achievement at 15.[32] Danish researchers found that kids who start school at six rather than five were much less likely, even five years later, to be diagnosed with hyperactivity and inattention.[33]

These findings were emphatic – if you started on the decoding before you had an underlying understanding of story, experience, sensation and emotion, then you became a worse reader. *And* you liked it less. Our attitudes to early learning were putting kids off for life.

Instead Hirsh-Pasek wanted kids to embrace the joy in learning and growing up. Apart from kids her other great love was music, and she often broke into song, especially on the phone to her grand-daughter. When her own children had been small she'd performed concerts and piano recitals to the delight of the neighbourhood kids and even her own. Hirsh-Pasek knew from experience what worked. In her book she suggested six Cs for modern learning: collaboration, communication, content, critical thinking, creative innovation and confidence. Truisms, I'd thought, but unlike much education policy drawn from real scientific evidence. If I was to take away one thing, she said, it was that 'from the earliest ages we learn *from people*'.

It was the same insight that had prompted a pair of suburban scientists to hit the 'record' button.

The World's Largest Home Video Collection

Deb Roy was dressed in black and still youthful when we met at MIT, flecks of grey hair the only evidence of 11 years of parenthood. Looking back, the Human Speechome Project seemed a quirk of turn of the millennium enthusiasm about artificial intelligence, a we-live-in-public commentary on the panoptic power of the internet that spawned Chat Roulette,[34] reminiscent of the heady days of Sixties psychology when they gave psychopaths LSD and locked them naked in communal rooms for days to talk it out.[35] The home videos had been gathering dust. 'I still have the whole collection,' he said. 'I'm waiting for his wedding day just to bore the hell out of everyone.' There were a few highlights. Two and a half million viewers had seen Grandpa snapping that first morning at home. Crowds had witnessed his son's first steps. But these were the only glimpses of footage that had ever been revealed outside the trusted core group.

'People ask, don't you worry about privacy? And I'm like, "You have no idea how boring everyday life is, especially looking back."'

In all, they'd captured 90,000 hours of video and 140,000 hours of audio during the experiment. The 200 terabytes of data added up to 85 per cent of the first three years of his son's life (and eighteen months of his little sister's) in glorious technicolour.

In a way it was also a great *lost* home video. With his team at MIT, Roy had developed new approaches to visualising and studying the data they'd captured: 'Space–Time Worms' rendered in abstract expressionist canvases the way people moved through the home over time; 'Social Hotspots', which showed two fragile 3D lines binding together in a tightly interwoven knot, were visual traces of tender moments in which parent and child came together to chat, learn or explore; 'Wordscapes' were snow-capped mountains ranged throughout the living room and kitchen, the highest peaks rising where particular words were most often heard. The tools had turned out to be fantastically lucrative as a means for analysing talk on Twitter. Roy and a graduate student had spent the decade building a new media company.

'If I could have cloned myself,' he said, less figuratively I guessed than others that used the phrase, 'we would have built on this in a more straight path.'

Roy *was* now back, and with a changed point of view. His new MIT group was called the Social Machines. He'd given up building robots to compete with humans and turned his attention instead to the augmentation of human learning. As we talked, Toco stood hunched and lifeless on a plinth in the conference room, a relic of some distant half-formed race. Roy had failed in his bet with Patel.

'I guess, putting on my AI hat, it was a humbling lesson,' he continued, 'a lesson of humility, of like, *holy shit*, there's a lot more here.'

He was no longer sure you could bring a robot up like a real human. Or if we should even try. It didn't seem there was much to gain by developing robots that took exactly one human childhood to become exactly like one young adult human. That's what people did. And that was before you got into imagination or emotions, identity or love, things that were impossible for Toco. Watching his son, Roy had been blown away by 'the incredible sophistication of what a language learner in the flesh actually looks like and does'. Infant humans didn't only regurgitate; we created, made new meaning, shared feelings.

'You're still learning, I'm still learning,' he went on. It wasn't decoding, as he'd originally thought, but something infinitely more continuous, complex and *social*. He was reading Helen Keller's autobiography to his kids and had been struck by her epiphany at understanding language for the first time. Deaf and blind after an illness in infancy, Keller was seven years old when she got it. 'Suddenly I felt a misty consciousness as of something forgotten,' she wrote, 'a thrill of returning thought; and somehow the mystery of language was revealed to me. I knew then that "w-a-t-e-r" meant the wonderful cool something that was flowing over my hand. That living word awakened my soul, gave it light, hope, joy, set it free! Everything had a name, and each name gave birth to a new thought. As we returned to the house every object which I touched seemed to quiver with life.'[36]

For Roy, it confirmed his revelation. The acquisition of language was embodied, emotive, subjective. It was the opposite of AI.

He was now seeking to build on this insight. Living with his kids, he'd been recognising 'more and more how important the social dimension is'. He was working on a project with Kathy Hirsh-Pasek. 'It's kind of obvious. You have to be a very peculiar type of linguist to believe that language has nothing to do with communication. Or that communication has nothing to do with social links.' It was a truth that was often lost from sight. 'There are such linguists walking around this campus,' he chuckled, looking out over MIT, 'or so I've been told.'

Roy was now researching at the 'intersection of language, technology and children'. AI wasn't going to lead the show, but he thought it might play a supporting role in transforming learning. There were two energy sources to tap: the analytic power of computers and latent human capacity 'at a hyper-local level'. He'd thought about how hard it had been to live up to the idealised view of a parent that he'd held in his head – and how this was harder still for those in poor communities. Everything at the youngest age depended on 'the socio-emotional context that the learner is embedded in'. This was and always would be established by parents who would feel stretched, short on time or ill-prepared. He wanted to give them the tools to succeed.

'The ability of this parent to make a difference for their kids is being blocked by certain things that technology can unblock,' he said. It was a little tech-utopian, but it was also similar to the ideas that drove Pen Green. He now imagined a 'three-way social network' of a parent, 'family learning coach' and child, supported by mobile-technology platforms in which kids played learning games. The data would analyse itself, telling you how the kid was progressing and sending prompts to both parent and coach. Roy didn't want to lose the 'essential human link', in this case between the coach and the parent. 'I don't think you can automate it, or want to.' Knowing a person had your back was more valuable than dialling Robonanny.

It was a crucial insight for how we thought about learning. A top artificial-intelligence expert, Roy had concluded that our future development lay in the exploration of our own capabilities. We had to start early and tech could help. Pen Green and the Temple Baby Lab hinted at how we might get started, but there was much more we could do. Language mattered. So did play. You had to learn socially from day one and you needed to feel loved. How could we accelerate all of that? Roy thought it best to follow what came naturally to us as humans.

During his talks, as well as showing his son's first steps, which took place in the hallway, Roy liked to play a 40-second audio clip of his son learning to say 'water' over a period of half a year. The audience heard 'ga-ga' repeated over and over, sung, whispered, inquisitively, demandingly, until gradually 'wuh' and 'tah' sounds crept in, disappeared again and reappeared, finally arriving triumphantly at 'water!' Roy had 2,000 of these paths parsed out from the data. It reduced people to awed tears to hear them. But even these Roy thought only revealed a small fraction of what it meant to learn. Roy was enchanted with another moment. He remembered above all the birth of his son's first word.

The first time he uttered something that wasn't just babble, Roy and his son had been sitting looking at pictures.

'He said "fah",' Roy explained, 'but he was actually clearly referring to a fish on the wall that we were both looking at.' He recalled every detail. 'The way I knew it was not just coincidence there was a fish and he said it,' he went on, 'was right after he looked at it and said it, he turned to me. And he just had this – it was like the cartoon-light-bulb-going-off kind of look. It was this "*Ah*, now I get it" kind of look. He's not even a year old, there's a conscious being – in the sense of being self-reflective.' He stumbled a little over the words. 'Not just getting it, but *realising he got it.*'

We were at our most open to learning in our earliest years. *Everything* was going in. Evolution still trumped technology. Rather than pushing in literacy and numeracy, it seemed we were better off working *with* our millions of years of beneficial adaptations, following what occurred naturally. 'We shall not cease from exploration,' wrote T. S. Eliot. Roy hadn't. His inquiring mind echoed that of the scientist in the crib. After journeying into the brains of robots, he was more deeply in awe of the born learning capacity of human infants. I didn't doubt Toco would one day return, but I thought in future he'd know his place. The most far-reaching discoveries were still to be made at the frontiers of our own minds. There would be no brain implant. Instead we had to start from a new understanding that we were uniquely *learning* animals. We needed a new science of learning that encompassed so much more than the drilling of times tables and memorisation of facts. 'And the end of all our exploring,' continued Eliot, 'will be to arrive where we started and know the place for the first time.'[37]

How could we maximise our brain power in the long term? What would it mean to make a science of learning? I knew just the place to turn to for an answer. The final stop in the first part of my journey would be a London school whose kids learned more in seven years than any others in the UK, and whose ethos went against everything technology was teaching us.

CHAPTER 3

Brain Gains

Going with the Flow

> I am the sum total of the parts I control
> directly.
>
> Daniel Dennett

Climbing the Mountain

The first sign that Ifrah Khan's new secondary school was a little out of the ordinary came on a spring day a decade ago when two 'very prim, very proper' men knocked at the door of the flat she shared with her parents and three siblings. 'Everyone was like who *are* these people?' she remembered. On her Hall Park Estate, close to the shisha bars and takeaways of London's Edgware Road, smart suits usually meant a wedding, or occasionally a court date. But these men had come to visit her. Their names were Mr Haimendorf and Mr Patterson, and after settling into the family sofa they pulled out a contract. Perched between her mum and dad on the other side of the coffee table, the ten-year-old Ifrah was spooked.

'No other head teachers were coming to homes and making us sign an agreement,' she told me over tea on a dark winter's evening in Yorkshire. Eight years had passed since the visit and Ifrah was in the first term of a law degree at Leeds University. She had mastered pasta with pesto via YouTube tutorials and FaceTiming Mum, and was surviving essay crises on a buffet of Chocolate Fingers, Pringles and Diet Coke. She was a born Londoner, who confessed she could 'talk for England', and her slight frame, white quilted coat and expertly applied makeup accompanied an original mind and steely determination. When she wasn't reading up on tort law or writing about the English legal system, she watched *Gossip Girl* and rated the town's nightlife as *crazy good*. But Ifrah was also focused. She'd missed just one lecture so far and was already plotting her career at a City law firm.

'I know where I am and I know how to get to where I want to be,' she explained, screening out Danny Dyer's *Who Do You Think You Are?* on the common room TV as her roommates watched. A George Foreman Grill oozed on the countertop. Ifrah knew how tough it was to succeed. Earlier, as we'd toured the lavish Liberty Building, home to the Law Faculty, she'd paused to take in the scene.

'I never actually thought that one day I would be walking around like these people,' she confided.

It was this thought that had prompted my visit. In the UK, where your family background and parental earnings still dictated your life chances, Ifrah ought never to have made it to Leeds. Growing up in one of London's poorest wards, her peers were destined for gangs, not Russell Group institutions. Her two brothers had been sent to prison for a gang-related crime. When her parents were also arrested, Ifrah was put into care. The summer before she left for university was the first time in eight years her whole family had been together. As a British-Pakistani Muslim girl the odds were stacked even further against her. 'The traditional route is to get married off, have kids and live life that way,' she explained. Her sister was a full-time mother. If it hadn't been for the two men that came to visit her that day, she would have been too. Though it was a good life for those that chose it, Ifrah wanted something different.

'From a very young age they engineer it into your brain that this is where you're gonna go,' she said, gesturing at the grand old university library and nearby Co-op. 'This is what you're gonna do.'

After the sofa summit her parents put pen to paper, and, with a shrug, so did she. 'I thought no more of it.' That July she was shocked to find herself in an intensive two-week summer programme with 60 other kids from Church Street council blocks. 'It was an intense army programme!' she complained. 'It wasn't normal.' On balmy

August days when she longed to ride her scooter and play out, she was expected to turn up to her new school at 8 a.m. each morning and stick around until 5 p.m. at night. She was miserable. 'I was a very naughty child,' she admitted. Her mum told her to keep at it. 'I had to wear a T-shirt that said "Work Hard, Be Nice". I'd come home and everyone would laugh at me.' When her mates asked what she was doing going to school in the summer, she'd tell them, 'Oh, it's KSA. I'm going to KSA.'

KSA stood for King Solomon Academy. It had been set up earlier that same year by one of the men in suits. Convinced that all kids were capable of learning more, he'd set out to prove it by opening a mixed two-form entry school for three- to 18-year-olds that would work exclusively with the poorest children, who were also the furthest behind. It would have 'a no-excuses, high-performing, university preparatory ethos', he told me, where pupils and staff would do whatever it took to succeed. Under the right conditions, he thought, you could hardwire success into the brain of every child. Ifrah's parents had heard about the school in an information session at her primary and signed her up. The two-week boot camp was step one in a remarkable journey that would eventually lead a cohort of the country's poorest kids right to the top of the performance charts. That's why I'd taken the train up to Leeds to meet her one grey November evening. The going would be hard, for kids *and* teachers. And it would reveal to us just how much our kids could learn. This chapter tells Ifrah's tale. It's a story about the potential of our brains and what we can do to help more kids reach it.

'I tell everyone all the time. Mr Haimendorf, he's the man,' said Ifrah, 'he's the man who changed the game.'

On the Limits of Intelligence

How much *could* we learn? At Walworth I'd been sure that class 10X4 were capable of achieving more. KSA was the final stop in my effort to think anew about our capacity to learn because there kids *had*. And where once we thought it was alright for kids to muddle along, today we knew that wouldn't cut it. Our evolving economy and a planet stretched to its limit demanded we use *all* of our human potential. For the three in four kids in poor countries failing to reach proficiency in reading, maths and science, or the three in ten in the developed world who missed out in one of those areas, I hoped KSA would offer some answers. In the past we had the excuse that some kids just weren't cut out to learn. Ifrah's own primary school results would have seen her written off with a low IQ. Schools were full of this bad science. I remember comparing notes with my mum when we were both teaching, following training on Howard Gardner's Theory of Multiple Intelligences. It said that all kids had a learning style – linguistic, spatial, musical, kinaesthetic – and that you should teach to it. The theory had been published in the early Eighties, disproved in the Nineties and enshrined in teacher-training syllabuses in the 2010s.

It had started with Francis Galton. In 1868 the polymath cousin of Charles Darwin set out to settle a 2,000-year-old debate about human intelligence. Were we the product of our environment, a *tabula rasa* etched by experience, or was our ability innate? Galton, a child prodigy who'd been disappointed aged five to find that none of his classmates had read the *Iliad*, observed that clever people often produced gifted offspring. It led him to a hypothesis, which he tested in the field and reported on in his first major work, *Hereditary Genius*. The clue was in the title. Intelligent parents produced smart kids.[1] The flames of his ideas were inadvertently fanned by Alfred Binet. At the start of the twentieth century, the French psychologist developed a test to evaluate a child's developmental stage, designed to tell which *enfants de la patrie* might need additional support in school. The set of challenges ranged from asking the child to follow a beam of light to having to recall seven random numbers, or find three rhymes for obedient.* The result gave the child's mental age. Though the Frenchman intended his tool to show simply the point a child had reached *so far* in her learning, it was seized on by an American psychologist with a different view of intelligence.

Lewis Terman (whose son, Frederick, would coincidentally go on to found Silicon Valley) was a eugenicist and fan of Galton's. He believed intelligence to be hereditary and saw in Binet's test a way of

---

* Deviant, ingredient, expedient.

sorting those of differing mental ability. During the First World War he used his Stanford–Binet scale to administer the first mass psychometric evaluation, rating 1.7 million soldiers from A to E in general intelligence. Alphas went to officer training, Epsilons to the infantry. This marked the birth of modern IQ tests, and the sticky assumption that intelligence is fixed. But actually that's not what IQ tells us. You generate a score by taking someone's mental age as diagnosed by the test, dividing by your chronological age and multiplying by 100. Simple. But misleading. IQ is a *relative* measure, a snapshot in time. We came to view it as a fixed label for life.[2]

Or, at least, we did. Until neuroscience let us peer more closely at the grey matter inside our skulls.

A Whole New Can of Slugs

The aplysia is a strange beast. At home in shallow tropical waters, it's a hermaphrodite marine mollusc the size of a guinea pig, which if attacked squirts out a toxic cloud of ammonia. The Greeks named it 'sea hare' thanks to its two antennae and mottled brown colour, but in reality it looks more like what it is, a sea slug. It's also the dunce of the animal kingdom. Whereas humans have 100 billion neurons in our brains, the aplysia has just 200,000. Because of this, the aplysia (specifically the *Aplysia californica*, or Californian sea slug) is a big hit among neuroscientists. Chief among them is the bow-tie-sporting Austrian-American Nobel Prize-winner Eric Kandel, who as a student at Harvard in the 1940s had, like almost everyone else, been deeply influenced by B. F. Skinner, the founding father of behaviourism. But where the old hand saw psychology strictly as the study of behaviour and believed thought (like free will) to be an illusion, Kandel, a more sensitive literature major and devotee of Freud, was intrigued by thought and set out to understand its neural basis.

What happens in our brains, he wondered, when we use our minds?

When a youthful Kandel locked eyes with an aplysia across a crowded lab one afternoon in Paris, it was love at first sight. The simplicity of the sea slug's neurology meant he could study individual neurons and synapses in its brain, while one of the creature's natural behaviours, a reflex that shut its gill when it was touched with a stimulus, made it an ideal subject for experimentation. In a ground-breaking study, Kandel found that after 40 or so gentle caresses, or 'non-damaging touches',* to the gill the in-built reflex would cease. More remarkably, the cessation of the reflex could be traced in the neural structure of the aplysia's brain. Not only were the changes to the aplysia's brain biochemical (the concentrations of neurotransmitters between the motor-neuron synapses went down), but they were also *anatomical* (the physical configuration of the axons and dendrites emerging from the neurons was *different*). 'We could see for the first time that the number of synapses in the brain is not fixed,' wrote Kandel in *In Search of Memory*, 'it changes with learning.'[3]

Kandel's findings were revolutionary, the first unequivocal proof of the brain's plasticity. Before, our neural architecture was believed to be imprinted at birth, unfolding over childhood like those little 'dinosaur egg' toys that bloom into a triceratops or T-rex when dunked in bowls of water.[4] This gave us our belief that intelligence was fixed. But Kandel showed that the brain's very anatomy – not just the signals running along the pathways, but the pathways *themselves* – were altered by experience. Though Galton's myth persists in classrooms, the changes that Kandel recorded in the aplysia brain demonstrated conclusively that intelligence was hugely malleable. Both nature *and* nurture play a role. What is true for sea slugs is true for us. Our genes establish certain guard-rails within the brain. Beyond that the growth of one's intelligence is an unending, individual process.[5]

For schools, the implications are as radical as Mr Haimendorf's wager. Grades don't set down the limits of kids' potential, but simply show where they've reached so far. Moreover, it signals that every child is different, their own unique neural circuitry an inimitable thumbprint of strengths and abilities. This neurodiversity is most apparent in those with diagnosed learning differences. As Kandel told Andrew Solomon, whose *Far From the Tree* is unsurpassable on difference and identity, 'if we can understand autism, we can

---

* Proving that, whatever the doubters say, romance lives.

understand the brain'.* We are only beginning to scratch the surface. Every individual learner is different. Current performance isn't destiny for anyone. KSA's challenge was a challenge we all shared. Aged 11, Ifrah and her classmates, through no fault of their own, hadn't yet made as much progress as other kids. KSA recognised this and was making a big bet that all of its students, regardless of their background, had enormous potential that hadn't yet been realised.

On the train back to London after meeting Ifrah, I wondered how they might. Was there some ingenious way we could tap into the full potential of our minds?

Hopelessly Hooked on Learning

The techno-utopians thought they had found the secret. 'The way we learn today is wrong,' wrote Peter Diamandis, one of the founders of Singularity University. An entrepreneur and space obsessive, he was responsible for the X-Prize, a multimillion-dollar fund that aspired to build 'a bridge to abundance for all' by solving the world's 'grandest challenges' through awakening the 'creative, entrepreneurial and inventive spirit' within everyone. It offered $10,000,000 grants to those who could make new technologies like passenger-carrying spaceships, TED-talk-giving AIs or mobile devices able to diagnose patients better than certified physicians. I'd missed him in the Valley, but knew he had his eye on schools. Technology streamlined human experiences. If we could make learning less boring and more user-friendly, we could turbo-charge our minds.

'Learning needs to be less like memorisation,' he continued, 'and more like Angry Birds.'[6] It was a departure from Kathy Hirsh-Pasek's

---

* *Far From the Tree* is a magisterial multi-award-winning volume on difference that everyone should read. Solomon treats the subject of difference with an authority and empathy that are beyond my capability. If that's your interest, please, please put down this book and pick up that one. (Then once you've read it, pick up this one again.) I know the topic is of huge importance, but it's not one I have space to explore in this book. Todd Rose's *The End of Average* is also a brilliant, informative read on the subject.

view, but I saw his point. Kids were hooked on tech in a way that was hard to match in the classroom.

I thought back to the Christmas break of my first year as a teacher. Things had slowly unravelled. I'd been promised a relatively easy ride in my first term, a grace period when the pupils would be wary of a newcomer. Kids would sit back and weigh me up. Was I strict? Enthusiastic? Inexperienced? Did I care? Where were the boundaries? *Were* there boundaries? Then the testing would begin. Failing to follow an instruction. Claiming work was too hard. Trying to duck out to the toilet. I prepared for the opening salvos by banking two weeks of lesson plans and carefully designing a behaviour system.

'Can I go toilet?' Dean had asked, materialising under my left armpit on the first day. My Year 9s were about to begin reading *The Boy in the Striped Pyjamas*.

'No, Dean, sit down,' I responded calmly and, I hoped, authoritatively. How had he got there?

Dean didn't move.

'*Please*, sir. I've got a note.'

'He has, sir, he's allowed to go,' piped up Maria. 'He's got *permission*.'

We'd been told by the head that no one was to leave the classroom during lessons under any circumstances.

'You have to wait to the end.'

Unmoved, Dean continued to gaze up at me. I'd felt suddenly powerless. I had no means to compel a thirteen-year-old boy to sit down. It was becoming a stand-off.

'Let him,' chirped Harry.

'*Please*, sir,' pleaded Dean.

'Be quick,' I said, caving in. No harm in it. Dean had bustled off looking pleased. We returned to our reading.

That had been the first chink, I realised, as I sipped tea in my parents' lounge, looking disconsolately at the tower of marking on the coffee table. It was like the New York mayor, Rudy Giuliani, had said about crime: if you didn't sweat the small stuff – the broken windows – the big stuff mushroomed out. It had. The books were a testament to a disengaged class. A few were lined with neat paragraphs and careful presentation, but most seemed to carry obtuse messages from a dysfunctional civilisation. Interpreting the scratched marks, unfinished sentences, cartoonish pictograms, notes to self – *this is so boring* – I collapsed back on the sofa. One message was clear. The kids hadn't learned much. Not the wide-eyed Year 7s. Not the savvy Year 9s. Not the bottom-set Year 11s.

'You've got to capture their attention,' my tutor had told me after observing a particularly bad class before the half-term break. *This* was where Diamandis came in. In our age of boundless information, the most priceless commodity was our attention.[7] The genius of tech companies like Google and Facebook was their ability to capture it. In 2016 it was estimated that we checked our phones an average of 221 times daily. The four hours a day that British 16- to 24-year-olds spent online was shocking, but it fell well short of the five and a half hours of screen time invested by their American peers. More than 155 million Americans now played video games, spending a total of 3 billion hours *per week* controlling an online avatar in an imaginary world. We were helplessly in thrall to the screens in our hands.[8]

In 1998 an up-and-coming psychologist called B. J. Fogg opened the Persuasive Technology Lab at Stanford University. He had made his name with an experiment showing how human behaviour could be influenced by computers. In 'Silicon Sycophants: The Effects of Computers That Flatter', human test subjects working on tasks received different levels of encouragement from their computer collaborators: sincere praise, flattery or generic feedback.[9] The results revealed that even though the humans knew the encouragement was machine-generated, the flattery subjects tended to perform better on the task than the generic-feedback subjects and equal to the sincere-praise subjects. They also felt more cheerful about their experience and had more 'positive regard' for their robotic confreres.

'The effects of flattery from a computer,' Fogg concluded, 'can produce the same general effects as flattery from humans.'

In the world of interactive technologies, the news spread like cat-gifs across Facebook. The 'principles of reciprocity' that psychologists knew governed human relations (people return the treatment they've received) seemed also to hold for human–machine relations. Adding credence to the idea that computers could one day become social actors, it showed too how easily machines might manipulate us. 'We are suckers for flattery,' wrote Fogg. In an era when we were spending 40 minutes a day on Facebook alone it promised real power. Using these psychological insights, you could design apps to influence human behaviour. Fogg called his new field 'captology', not for its power to capture human attention but from an acronym: computers as persuasive technology. Predictably, he foresaw a utopian future in which educational software persuaded students to study for longer. Others saw dollar signs.

Behaviour design, as the field is now known, was not new. Its roots can be traced back to 1930, when B. F. Skinner had introduced a first rat into his 'operant conditioning chamber' while studying for his master's degree at Harvard. The Skinner Box, as it became known, was a small sealed container with a lever at one end that would release a pellet of food when pressed. On the first few goes the hungry rat would knock the lever by accident, releasing snacks haphazardly. But, after a while, it would pitter-patter straight up to the lever upon entering the box. The reward reinforced the behaviour. Skinner came to believe that the behaviour of all 'operants', whether rat, pigeon or human, might be programmed in this way. Environmental conditions, rather than conscious choices, caused behaviour. Free will was an illusion.[10]

In 1948, after the carnage of the Second World War and spurred on by his findings, Skinner published a novel. It imagined a utopian community – the eponymous *Walden Two* – guided by his new science to bring about social justice and well-being for all. In it, citizens did what came 'naturally' to them as a consequence of the tight design of the community by 'planners', like a grown-up Pen Green, minus the feelings. Individual agency would be surrendered to the closely administered influence of positive and negative reinforcement, terms familiar today to all teachers. Many other Skinnerisms survived in psychology, the media, advertising and schools, but the realisation of his utopia was hamstrung by the media of the era. It would require a totalitarian government to bring it about. Until today.

Now, in our pockets, we each carried a digital Skinner Box. Couldn't *it* be programmed to shape our behaviour?

'Yes, this can be a scary topic,' read Fogg's Stanford homepage. *No shit.* He promised a future of 'machines designed to influence human beliefs and behaviours'. For good, naturally. 'We believe that, much like human persuaders, persuasive technologies can bring about positive changes in many domains, including health, business, safety and education,' added his website. They could. Students who studied at his lab had gone on to found Instagram, design Google's user experience and consult for tech firms in how to build habit-forming products. The technologies were persuasive, but were they enacting *positive* changes? 'We need to make kids as addicted to learning as they are to gaming,' Peter Diamandis said, implying that harnessing the attention-grabbing power of our hand-held devices could fuel a learning revolution. I wasn't so sure.

Following the crisis class, we had put together an action plan. I was spending so long failing to get the class to behave that the kids who wanted to get started had nothing to engage in. This would change. My tutor proposed a regime of simple starter activities. No handing out books. No requirement that everyone begin at the same time. No need for an explanation. I would have a simple exercise awaiting kids on their desks or the whiteboard when they entered. They would come in silently and get on with it. I'd use praise to highlight those that were getting it right and give out merits rather than warnings. The currency of learning was attention and I was going to trade in it.

The Year 9 class was transformed. They listened in awed silence to *The Boy in the Striped Pyjamas,* read pages of text and answered question sheets on the previous day's material. Every lesson the short tasks were waiting. And every lesson 28 kids would come in and get on with it, giving me time to deal with the Deans and Marias. Best of all were word searches. They were catnip to the kids. Thirteen-year-olds could spend five, ten, fifteen minutes finding the words, and they'd even plead for more time to find the last one or two when I tried to move them along. They were captivated.

But as I started to go through the exercise books that Christmas, another question had stalked me. Yes, they were paying attention. They were doing activities instead of asking to go to the toilet. But were they *learning*? Their books suggested not. Leafing through the pages, I realised they'd hardly learned a thing. Word tables ranked alongside firing cartoon canaries into tottering towers of pigs as a learning experience. You could lose yourself in the simple, repetitive tasks, but you weren't better off for it. If the radical power that machines could exert over our attention was to unleash a revolution, we had to better understand the true nature of learning. That's what had led me to Ifrah in Leeds. For seven years at KSA, she and her classmates had learned more than almost any other group of kids in the UK. Their exam results had proved it. This wasn't some vague hunch, but a remarkable, proven story of real-life success. I went along to her school to find out more.

Doing Whatever It Takes

King Solomon Academy nestled incognito amid the chichi galleries and shady side streets just off London's Marylebone Road. At 8 a.m. sharp on a cold November morning the school gates on Penfold Street locked shut and Mr Haimendorf briefed the staff. Rumours of a blizzard swirled, but there were no latecomers. Fifty youthful teachers ate Coco-Pops and drank instant coffee as he sped through notices. Year 7, 8 and 9 behaviour. Black History Month. Holiday homework. A Year 8 football tournament. The Science Museum outing. It was purposeful, more business-like than school-like. After five minutes and 33 seconds the meeting was closed. We jumped up, coffees in hand and rolled out to greet the students.

Today KSA ranks among the top handful of schools in the whole of the UK.* In 2015 its students achieved the top GCSE results for a comprehensive school in the country with 95 per cent attaining

---

* There are myriad ways to rank schools. In pure GCSE performance, St Paul's Girls tops the most recent charts just ahead of Westminster School, with nine of the top ten places taken by other private schools. The highest-performing state schools are all grammars, and therefore selective. More interesting are the tables that rank the non-selective schools in overall performance and Progress 8. These are the truer measures of the quality of the school, rather than the quality of the student intake. Thomas Telford in Shropshire regularly triumphs in overall outcomes, while on the Progress 8 measure, which looks at value added, Tauheedul Islam Girls' Grammar School in Blackburn is a regular top performer, as is KSA.

five A*–Cs including English and Maths. On the Progress 8 score, which measures how much pupils have learned in their five years at school by subtracting starting levels from finishing grades, it has been among the top few schools nationally since 2016, when the measure was introduced. KSA students joined aged 11 as future foot soldiers, a few years behind their expected reading and writing levels, but graduated at 18 as officer material. Like Ifrah, nine in ten leavers went on to university, despite the majority of kids qualifying for free school meals.[11] The school's success was an irrefutable testament to the malleability of human intelligence – and a head teacher's dream.

An Oxford science graduate, Max Haimendorf had known that the conveyor belt into the City was not for him. He'd taught in Tonga during a gap year and then signed up to the first cohort of Teach First, a programme that promised to fast-track top graduates into inner-city schools. 'It had this catchy "learning to lead" thing,' he told me. As a novice teacher at a west London secondary school he excelled, quickly demanding a form group, becoming head of year and learning to 'sweat the small stuff', leaving no top button undone, no litter unpicked. Where well-meaning colleagues patiently explained that he shouldn't push too hard as 'not all of the kids in this school are able to do their homework', Haimendorf reacted against the low expectations ingrained in the system, forming a fundamental belief in the potential of all kids to succeed, given the right support, expectations and environment.

After a year-long stint working on special projects at Teach First, he took a brief detour into financial services, working as 'an Excel-monkey' before quitting. On Saturday mornings he had begun meeting with a group of like-minded education radicals, outlining a shared vision for the ideal school over cappuccinos and croissants. They realised that if they were to make good on their dream, one of them would have to go full-time. Among the coffee revolutionaries, one candidate stood out and Haimendorf soon found himself being approached by the ambitious founding board members of a large academy chain, who decided to entrust him with millions of pounds of public funds – and the lives of hundreds of kids – so that he could bring the school to life. The day KSA opened, he became the UK's youngest head teacher, a fresh-faced 28-year-old with bold ambitions. It was, he says, 'a jump into the abyss'.

'You're trying to make every second count, every interaction count,' he explained from behind the desk in his office when we caught up later in the day. His bookshelves were lined with well-thumbed copies of big-ideas books like *Black Box Thinking*, *Freakonomics* and *Driven by Data*, which showed he was always on the lookout for new ways to improve. Now in his mid-thirties, he was a little greyer round the temples, tired with the long days and a new baby at home. Yet his ambition for KSA's kids was undimmed, his suit and tie as prim and proper as Ifrah remembered. Everything he did flowed from a simple aim: every student would make it to a good university.

It started with time. If you believed in the malleability of minds, more time meant more learning. The school day stretched from 7.25 a.m. to 5 p.m. to give the kids extra hours. Two further weeks of study were added in the summer holidays and two hours of homework expected each evening. Next was the quality of relationships. Whereas in typical secondary schools teachers interacted with 400 kids a week ('They hardly know people's names, let alone how to help them learn,' said Haimendorf), at KSA there were just 60 kids in each year. Teachers were assigned to a particular year group, which saved on planning time and created a multiplier effect. They really knew the kids. Finally, there was an emphasis on depth before breadth, 'nailing English and Maths before everything else'.

Kids wouldn't choose from many options, but the ones that were available to them, they'd *smash*.

The learning regime started with Year 7 boot camp. 'You've got to create that culture,' he explained. In the first summer they engineered a bubble around the kids, keeping them off the street and away from distractions, like elite athletes at a pre-games training camp. They openly conditioned them, making it 'normal during that two-week period to become highly compliant, get it right behaviourally all the time, to complete all your work, to be praised consistently and frequently'. The school was developing kids' learning muscles, strengthening their focus, ability to pay attention and perseverance. He wanted a tightly knit unit, a 'team and family', with their own identity, good habits and model behaviours. After a fortnight it would be 'cool and normal to praise your peers, cool and fun to do well at school'.

Strange rituals were introduced, clicks instead of claps, silent walking in the corridors, a centralised homework hand-in system, transparent pencil cases. No detail was too small to sweat. The school came under fire for its strict discipline. 'No excuses, at all, whatsoever,' Ifrah had told me solemnly. 'You're either wrong or you're right. You're either good or you're bad. There's a line and you don't cross it. And if you cross, it's game over.' Haimendorf was the Alex Ferguson of school leaders, holding the warm–strict line at all times. 'It's caring enough to believe that every child can meet the expectation,' he explained. 'Holding them to it in every situation is the right starting point.' Everything was thoughtfully tailored to mould kids' habits and maximise their learning time. The university dream kept them honest, meaning they couldn't obsess only over exam results. They had to raise aspirations, build character and learn beyond the confines of the GCSE. They did corny things, like name every tutor group after a university (Ifrah's best friend, Mey, was now studying International Relations at Manchester University after being in Manchester tutor group) and run sessions called 'Planet KSA'. Here kids googled university courses and role-played what they might do if at uni they came across someone passed out drunk. They also did life-changing things. Every student spent four hours a week learning a musical instrument, playing in an orchestra that had toured to Vienna and Paris. Every student attended a one-week residential each year at a top university.

'I remember Oxford very clearly,' Ifrah told me, 'I remember feeling so blessed.'

Kids absorbed the culture of high expectations by osmosis. They were so high that the year Ifrah and Mey took their GCSEs Haimendorf was disappointed with the stellar results. 'I made the mistake of articulating my disappointment to the people around me,' he confided. The staff who had done so much to achieve that success didn't appreciate it. And it was the people that counted. The school's success was based on visionary governance, great teaching and a nurturing relationship-based education. Every action by every person in the school environment, however small, was influencing the way the kids developed.

'Your brain changes,' wrote cognitive scientist Steven Pinker, 'when you are introduced to a new person, when you hear a bit of gossip, when you watch the Oscars, when you polish your golf stroke – in short, whenever an experience leaves a trace in the mind.'[12] It seemed that Haimendorf had set out to ensure that every single experience the kids had at KSA, including lunchbreak, where kids ate for half an hour in family style in groups of six, was carefully designed to leave the *right* trace in their minds. Was that the secret to maximising learning?

The Attention Game

Eric Kandel's sea slug love affair lasted a lifetime. For him the search for memory was an investigation into *why* certain experiences left deeper traces than others. The change he first observed in aplysia neurology was a transformation of implicit memory, which governed performance, habits and reflexes, into a form of long-term memory. The sea slug came to 'remember' not to shut its gill. The genius of our human brains was that as well as holding implicit memories, we could also create and store explicit ones, of people, places, events or ideas. Kandel labelled these 'complex' memories. They resulted from the same synaptic strengthening as implicit memories, altering the same neural pathways, both biochemically and anatomically. But they also seemed to differ in a crucial way, undergoing a process Kandel called 'system consolidation'.[13]

When we think about something – really think about it – we engage our working memory. It's the stage of consciousness. Capacity to tread the boards is limited to about seven explicit items. That's why it's difficult, without a mnemonic, to recall more than seven random numbers in a row. When we think, we call long-term memories from the wings and hold them on the stage next to objects in our immediate experience, so that sights, sounds, smells, tastes or touches can perform for us in front of our thinking mind. Though the experience of consciousness remains one of our great mysteries, we now know that conscious thought depends not on reflexes, like the gills of the aplysia or workings of our own lungs, but on a deliberate, willed effort. We know it as paying attention.

'The essential achievement of the will is to attend to a difficult object and hold it fast before the mind,' wrote William James.[14] Neuroscience now revealed this as an observable neurological state. Fired up through an act of will, neurons in the frontal lobes of the cerebral cortex send signals to the midbrain to begin producing the neurotransmitter dopamine. This enables us to turn down the volume on those things in our field of perception that aren't at that moment important to our enterprise, known as 'suppression', and to turn it up on those that are, called 'enhancement'. Axons from these cortex neurons meanwhile reach down into the hippocampus (deep in the brain, it's akin to the conductor of the cerebral orchestra), which then jump-starts the consolidation of explicit memory. So when we attend to something, whether by free will, a burst of emotion, under coercion or by finding meaning in it, we hugely increase our chances of remembering it.

Perhaps the type of attention that led to learning *required* willpower; unlike Angry Birds, which undermined it. You had to make a concerted effort to focus on those things you wanted to retain. I suspected it was impossible to game that.

Britain's Brightest Student

'The thing I think about most is attention,' said Daisy Christodoulou. We were sitting in an airy fourth-floor office of Ark, an academy chain responsible for supporting 35 UK schools, including KSA. 'That's why Facebook is a billion-dollar company.' In her early thirties and dressed in a smart black jacket and bookish glasses, she looked like a genius, which by some measures she was. A cult figure dubbed 'Britain's Brightest Student' after captaining a winning Warwick team on *University Challenge*,* she had gained a large global following when her book *Seven Myths about Education* dared to argue against the orthodoxy of twenty-first-century learning and in favour of a knowledge-rich curriculum. She was Ark's assessment guru.[15]

---

* You can see her performance on YouTube. It's extraordinary.

'In a world where information becomes cost-free, what then becomes in short supply?' she asked. Christodoulou was fingers-on-buzzers fast. Growing up a West Ham fan and cricket nut in London's East End, she won a scholarship to a prestigious private school for girls, where she continued to add to her encyclopedic knowledge. As we spoke she referenced everything from Hannah Arendt on totalitarianism to the techniques of FC Barcelona's La Masia training academy. It was exhilarating. She believed in a dream of human flourishing, a revolution in learning akin to that of the English Renaissance, when in the sixteenth century the sons of glovers, like Shakespeare, had first gone to grammar schools, transforming our culture for ever. Today she believed this was possible for everyone. 'There's so much human potential that we could realise.' The key lay in establishing a learning science, based on insights and techniques that had been proven to work. And the science suggested we still had to learn knowledge.[16]

I'd come to see Christodoulou because the central ideas in her book were that memory, whatever Peter Diamandis thought, was the true basis of learning, that it required consistent conscious effort to achieve, and that it was built on a bedrock of knowledge. Our system was ill set up for it. 'When one looks at the scientific evidence about how the brain learns and at the design of our education system,' she'd written in her book, 'one is forced to conclude that the system *actively retards education*.' The tech-utopians were right about attention, but they were wrong about learning. 'What you *think* about is what you remember. What you remember is what you learn,' she continued. The question then became: 'What are people thinking about in lessons?'

There was a cumulative effect to your cognitive development. It was a fallacy to believe we could develop general skills like 'critical thinking', 'problem solving' or even 'reading' in a vacuum. These always depended on a solid foundation of domain-specific knowledge. She referenced the fourth-grade slump, a well-documented phenomenon whereby kids from poorer backgrounds began to fall behind their wealthier peers, as reading became less about decoding words and more about understanding content. Even now, she rated herself a great reader when it came to the newspaper, the *Wisden*

Cricketers' Almanack, a historical biography or research on the science of learning, but a novice when reviewing the content of a colleague's PhD thesis on stomatin-like proteins, even though she could decode the words.

'Memory is the residue of thought,' she said, quoting cognitive scientist Daniel Willingham. His book, *Why Don't Students Like School*, was the bible on the use of brain science in education.

'Most of the time what we do is what we do most of the time,' he had written. Our brains were 'not designed for thought, but for the avoidance of thought'. In everyday life, say driving a car along a familiar road or playing a game of Angry Birds, we weren't *thinking*, but rather acting on memory. A very important element of learning was therefore the process of how you paid attention to something, thought about it and thus ended up with it stored. 'Things can't get into long-term memory unless they have first been in working memory,' he wrote. 'Usually, paying attention to a thing conscientiously means it will go in long-term memory.' You couldn't learn something you didn't pay attention to.[17] Yet the process of paying attention to something was complex, and not always under our control. It could be enhanced, according to Willingham, in a few proven ways: things that created an emotional reaction were much more likely to be remembered; repetition helped a little; wanting to remember didn't help that much; reflecting on meaning had a positive effect, such as knowing where something fitted in a story or schema, whether personal or general.

'A teacher's goal,' he wrote, 'should almost always be to get students to think about meaning.'[18]

'For me, one of the few limitations on learning is time,' Christodoulou said. 'The question becomes, how can we most efficiently schedule that learning so that people and adults can learn it efficiently?' She saw in the inner workings of computers a methodology for humans to follow. Complex skills could be broken down into simple steps and discrete building blocks. We could then master each of those in turn. 'Look at the amount of things a computer can do,' she said, 'I'm not saying that they can do everything, but they can do a lot.' On one level, it did seem like our minds *were* a little like machines, or could at least mimic them. There was something of this in KSA's approach. They maximised kids' learning time, kept them highly focused and fuelled them with carefully chunked reading, writing, maths and knowledge. Ifrah had said they 'engineered success' into her brain. Was that possible?

The Struggle is the Point

In a warm first-floor classroom at KSA, the 30 kids of Year 8 Goldsmith's Form (after the university) were getting going. By 8.29 a.m., about twenty minutes in, I'd written, 'these kids believe in a dream, of university, hard work and learning. So do I.' Twelve-year-olds were a notoriously tough crowd, but one by one they'd entered the room, sat down and pulled out books to read: *Skulduggery Pleasant*, *Percy Jackson*, *Dork Diaries* and *Flirty Dancing*. The form tutor maintained a close watch. 'Lovely, well done, Iman,' she said. 'Thank you, Ola.' 'Unless you're talking to an adult right now, you should be just reading your book.' Violins and cellos were piled in the corner. Homework journals had been checked.

'We are the ones we have been waiting for,' read a slogan on the classroom wall.

'5. 4. 3. 2. 1,' counted down Ms Harvey. *Every second counts!* 'Hands on desk. Eyes on board.' She waited. 'Two more people hands-free. Eyes on the board.'

At 8.31 the room erupted with the sound of 30 twelve-year-olds furiously clicking their fingers. Goldsmiths was on 96.9 per cent attendance. The form that did best that term would win a 'Classroom Multiplex' screening of *Johnny English*, popcorn included. They loved it. I could almost *see* the bubble Haimendorf told me they built around the kids. By 8.40 she'd departed the room, wishing the Year 8 girls' football team well ('Let's try to actually *win*!') and making way for maths. Books were swiped from bags and pens uncapped as the 'Do Now' was projected on the whiteboard. Every kid got on with it. The 'Do Nows' were a feature of every class, along with extension activities, bonus questions, exit tickets and clicks. All part of the KSA system. Fifteen minutes later, a one-minute warning was given and at 8.45 the teacher cut in.

'Pen in the air in 1. 2. 3. Thank you.' She flipped on to the next slide of her PowerPoint. As I furiously scribbled notes I noticed that none of the kids were asking her to slow down, or to wait. It was unheard of. I wrote: 'This is the equation: maximum learning time.' The teacher was a conductor, skilled at manipulating the kids' attention. If she wanted them to heed her, she clapped a quick seven-beat rhythm and the kids responded with two claps of their own. She chunked time into carefully managed seconds, ensuring they remained focused. 'You've got one minute thirty,' she said, introducing an activity. Motivation stayed high. The kids were going at it, learning to represent inequalities on a number line. A few quavered under the strain, but there was no gazing out of the window, no surreptitious passing of notes. They worked conscientiously throughout. At the end of the class, the teacher pulled up a final slide to award a MAPP Score, grading the collective for each of Mindful, Achieving, Professional and Prepared. Full marks I thought. But, no, they were docked a mark for wasting a moment getting hands-free and another for *silent means silent*.

I thought I'd known high expectations. This was something else. And, for the kids, it could be tough going.

'All kids don't like school,' Tarek in Year 8 revealed to me at break-time. 'It's a normal thing. But when you give them a test, they're gonna realise, oh, I am so thankful I went to that school. The teachers were mean, but look at me now. I'm a doctor or surgeon.'

It was a struggle to make it through KSA, but it was worth it. And I wondered if it was also the point. Sihana agreed. Her first impression was that 'It was a bit strict, and hard', but she knew it was good for her. 'Some of us don't really like the teachers,' she said conspiratorially. Both kids were conflicted about the rigours of the experience. Like eating their greens, they knew it was good for them, but it wasn't always fun. Occasionally the kids felt a little like cogs, grinding away towards their future. But then I thought again of Ifrah. In Leeds she had convinced me it was worth it.

'Uni, uni, uni,' she told me. It was all the adults talked about. She could still recall the universities that her teachers had attended and the day in Year 7 when Mr Reddy asked her what she wanted to be. 'I was *this* small and I didn't have any teeth.' She'd said, 'A lawyer!' But it had been a struggle, particularly after she'd gone into care.

Like Sihana and Tarek, for years she'd not had an easy time with the high expectations of the school. 'I just wanted to break walls and faces, I was so angry.' But there were no excuses, 'even if someone died'. The teachers would support her through it, but she wasn't allowed to be silly. The memories were ingrained. 'The word demerit just rings a million bells in my mind.' Everything was designed to keep her focused.

Kids at KSA *had* been conditioned to do well. But they were also in control, were being pushed to *think*. Real learning felt less like a user-friendly experience, more like a struggle.

The Paradox of the Guided User

'I think you have to be careful how stupid you make your user,' Christof van Nimwegen told me over Skype from his home office in Holland. A fast-talking psychologist of human–machine interfaces at the University of Utrecht, he had the linguist's ear for a one-liner. 'Money, ain't the damn thing funny!' he trilled. 'Grumpy. Bastard. Amsterdam,' he mock-dubbed himself. Computers, software and the internet fascinated him, particularly 'the trade-off of what I do and what the machine does'. He saw tremendous power, magic even, in technology, particularly screens. But he believed in the continued importance of human agency and had directed his research towards the division of labour between *us* and *them*. He'd made his name with a paper, *The Paradox of the Guided User*.[19]

Wanting to understand the effect of computer assistance on human performance, van Nimwegen devised a series of experiments in which he sat two groups of subjects in front of computers and asked them to solve increasingly difficult logic problems. In one condition an 'externalisation' group received helpful on-screen hints from the software about what to do, while in the 'internalisation' condition subjects were left to figure things out entirely on their own. He hypothesised that on-screen hints would help users learn the rules of the game more quickly and that they would perform better. But they didn't. Paradoxically, the results showed that although users in the externalisation condition answered the earlier problems more easily, as they progressed through the game and the hints ceased they were actually less well equipped to succeed, performing worse in the long run and being more likely to give up. When van Nimwegen tested users eight months later, he saw that the differences persisted.

Users who hadn't received help had derived a lasting cognitive benefit from working out the game for themselves.

When you talk about learning, he explained, if you sensed deep down that help was at hand, your brain became 'extremely lazy'. When the brain could take a short-cut it *did*. But, in doing so, it failed to lay down the cognitive architecture necessary to deepen its intelligence. A study of the effect of teaching assistants in the UK had found that, when controlling for all other factors, kids with the support of a teaching assistant actually made *less* progress than equivalent kids with no individual help. Without the right training, teaching assistants became kids' short-circuits around the necessary struggle of learning.[20] In the situation where 'we don't think any more,' said van Nimwegen, we risked becoming stupid. Technology, rather than augmenting our intelligence, limited it. It was another argument for continuing to bother with spellings and multiplication.

'There are many examples of people driving into the canal by using their navigation system,' he explained. In his *From Bacteria to Bach and Back*, the philosopher Daniel Dennett voiced a similar fear that people were beginning to overestimate the intelligence of their artefacts and becoming over-reliant on them, thereby threatening the institutions of human knowledge and intelligence.[21] Dr Lisanne Bainbridge at UCL called this the 'paradox of automation'.[22] Technological assistance undermined human intelligence (the software did all the thinking), while at the same time placing an even higher premium on human ingenuity (if something went wrong with the software, results would be catastrophic *unless a human intervened*). A learning theorist, Elizabeth Bjork, suggested that instead we should embrace 'desirable difficulties'. The friction of a hindrance, as long as it was not too great, resulted in a richer learning experience.[23]

'We want everything,' added van Nimwegen. 'Everything should be automatic. But one thing we cannot do is put knowledge in our minds, like you put a memory stick in a computer with information on there. You cannot do that with humans.' He paused. 'Luckily.' Learning had to be hard. It took time. If you weren't finding it difficult, you probably weren't learning. Van Nimwegen himself was contributing to a growing effort to build technology 'with a hiccup'. You added rough edges to the user experience, so that the brain wrestled with them and grew as a result. It was the difference between the enlarged hippocampus of traditional London cabbies and the presumably regular-sized ones of Uber drivers.*[24] 'If you want to be smart it takes solid work. You have to read, you have to contemplate, you have to discuss, you have to be present, you have to open your mouth. We should be aiming much more for what we can do with technology to make it richer, rather than to make it easier or faster.' If learning couldn't be easy, motivation had to play a big part.

The Difficulty with Desirable Difficulty

Back in California, B. J. Fogg had developed a model of human behaviour. Called B=mat (behaviour equals motivation plus ability plus a trigger), it suggested that our actions depended on all of these things coming together at the same instant. On the graph he used to present, the x-axis showed ability, ranging from 'hard to do' to 'easy to do'. The y-axis tracked motivation, from low to high. A concave 'action line' descended from top left to bottom right. Action, thought Fogg, happened only when motivation, ability and a trigger came together at the same moment and in the right balance. A 'hard to do' task could only be achieved, he thought, if motivation was high and the trigger was well timed. If a task was easy to do, then a trigger worked even when your motivation was low. The idea was to 'put hot triggers in front of motivated people'. Fogg believed his model, combined with algorithms that could help us understand when our motivations for different things were high, could help bring about world peace within 30 years.[25]

---

* One of the most famous 'proofs' of the malleability of the human brain came from a study of London's black cab drivers. To qualify for the job, cabbies first have to master the 'Knowledge', memorising 25,000 London streets, 20,000 landmarks and 320 basic routes. Researchers found that the acquisition of this mental atlas resulted in an observable enlargement of the drivers' posterior hippocampus, the part of the brain that deals with spatial representation. Uber drivers outsource this knowledge to their smartphones.

I was more sceptical. Fogg's behavioural insights were above all fuelling the attention-harvesting practices of Silicon Valley's app-makers. His former student Nir Eyal, who'd worked at Instagram, had written *How to Build Habit-Forming Products*. Another, Tristan Harris, had worked for Google, heading up their operation to op-timise the user experience, which would keep you on the site. The 'attention economy', he now warned, was engaged in a 'race to the bottom of the brainstem', in which behavioural-design specialists made their apps more 'user-friendly' and thereby more addictive, 'pushing us all to spend time in ways which we recognise as un-productive' and even meaningless, but which we are powerless to resist.[26]

Professor Natasha Dow-Schüll has shown just how much this was true in a study of slot machines in Las Vegas.[27] Carefully cali-brated digital slot machines – these days the casinos' biggest earners – transported players into a trance-like state, known as the 'machine zone'. Brought on by the subtly varied rewards, near misses and other psychological tricks embedded in the software, it was akin to a flow state and highly addictive. But where flow could be 'life affirming, restorative, and enriching' – a state of 'optimal human experience', repeat machine gamblers 'experience a flow that was depleting, entrapping, and associated with a loss of autonomy'.[28] The experience was *the same* whether you were performing Rach-maninoff's Second Piano Concerto or racking up points on Candy Crush. The difference was one of outcome.

Learning was affirming. It *had* to be difficult. If you were starting behind your peers, as Ifrah and her friends had been, then it would inevitably be even tougher. 'You don't come to school for fun and games,' she'd told me. 'You come to school to learn.' Ifrah's time at KSA had little to do with the bottom of the brainstem. It was all about the frontal cortex. It had been beset with challenges. Almost every instant of her seven years at school had fallen into Fogg's 'hard to do' category. From 11 to 18 she had made exceptional progress, more than almost any other kid in the UK. On average, her class-mates in the school had learned more from Year 7 to Year 11 than hundreds of thousands of others. It hadn't been a simple question of conditioning, but one of struggle. KSA was one long hiccup.

'You can't get people to do something they don't want to do,' Fogg told me. It was the central conceit of his model. But KSA had motivated students so that they *did* want to do things, helped them keep going even when it was hard. Fogg's long-term solution was to use technology to read our behaviours to assign tasks to people in the moments of highest motivation. But he conceded that there could be some tasks for which someone may never be intrinsically motivated. Algebra. Sports. Spelling. KSA might have used some of B. F. Skinner's insights around rewards and punishments to establish the school's culture at the outset, but then they'd built relationships, fuelled the kids' dreams, involved them in sports and music, done whatever they could to keep their motivation high and their focus sharp. I was struck by how motivated Ifrah was. Her dream of being a lawyer drove her on. She could already picture the glass-walled office with her name engraved on the door.

'Learning to think,' said David Foster Wallace in his famous 'This is Water' commencement speech at Kenyon College, 'really means learning how to exercise some control over *how* and *what* you think. It means being conscious and aware enough to *choose* how you construct meaning from experience.' If you weren't exercising some control, you weren't learning, at least not in an affirming sense. Learning couldn't be like Angry Birds. We had to be wary of seeking to improve ourselves by upgrading our technology. It didn't work. *You* had to be in control. If we became overly reliant on our machines, we risked eroding our own human intelligence. The ultimate aspiration of learning was to be able to weigh, to consider, to choose. We had to ask ourselves, were we learning to use the tools of today or were they learning to use us?

Our neural plasticity allowed a mechanistic view of the mind. But that view seemed limited. 'Every act of perception,' wrote biologist Gerald Edelman, 'is to some degree an act of creation, and every act of memory is to some degree an act of imagination.'[29] Our intelligence was *alive*. Human brains were unruly, infinite. We fed them through experience, learned to master our attention and focus. Their evolutionary genius was precisely that they adapted continually, organically to the environment and culture they were steeped in, or the increasingly complex tools at our disposal. Slowly, we were learning to engineer them. We could raise test scores, improve academic out-comes. New sciences of acquiring knowledge, learning to read or do maths, were emerging. KSA and Rocketship had shown me that. But this was to see our intelligence only partially. Our brains could empathise, collaborate and imagine. They would never fully comply.

In the next part of my journey, I planned to investigate further the capabilities our kids needed to succeed in our unpredictable times. I was sure our potential was greater than we realised, but I still wasn't clear what we should apply it to. KSA bet that kids should learn about academic subjects, music, self-motivation, team and family. The tech-prophets thought coding, creativity and complex commu-nication. Was education an end in itself, or did it mean mastering skills to succeed in work, or life? To form my own view, I'd journey next to Paris, where I'd heard of a futuristic coding school rewriting the rules of technical education, then on to Finland to see how we might learn to co-operate in shaping a happier, more sustainable future. It was clear our brains could function like machines, but were capable of more. If we failed to realise our full potential, we'd be left not only having failed hundreds of millions of kids, but also, as Foster Wallace put it, with 'the constant gnawing sense of having had and lost some infinite thing'.

PART II

DOING
BETTER

CHAPTER 4

Just Do It

Cradle to Career

> Live as if you were to die tomorrow.
> Learn as if you were to live for ever.
>
> Mahatma Gandhi

Don't Stop till You Get Enough

Lilas Merbouche was fed up. Barely 21, she made ends meet as a checkout assistant in a pet shop in Asnières-sur-Seine, a sprawling banlieue in north-west Paris. She was taking any opportunity going, 'really shitty jobs', she told me. The worst was event hostessing at the Stade de France, which was casual, piecemeal work. You never knew when they wanted you and when they did it was miserable, standing up all day, 'smiling all the time'. Still, she was fortunate to have work at all. Youth unemployment would hit an all-time peak that year in France. She dreaded the life unspooling ahead of her, dead on her feet by 5 p.m., scraping by on a few euros until payday. She wanted more; to find her purpose, to write her own story. There *had* to be more to life than dog bowls and budgerigars.

'I dreamed of a job where I could make a living,' she explained. But school had been a washout. 'It was a catastrophe.' She'd left the lycée without a single qualification.

'There are people who are made for school,' she said, unable to recall a single highlight, 'and then there are people like me.' She laughed at the memory, or my ill-pronounced French, then stopped.

'I really suffered.'

Lilas felt *nul* in class and spent her time bunking off. Lessons were irrelevant to her, rooted deep in a distant past. Aside from a single maths teacher in her first year of lycée, who'd had a knack for creating engaging lessons, the adults had bored her. 'Oh là là,' she exclaimed, when I asked what she'd change. '*Everything!* For all the years I was at school, I felt worthless.' Her older sister was a golden child, graduating with her *bac*, the French school graduation certificate, and heading off to the prestigious SUPINFO, an international institute of computer sciences in central Paris. Lilas knew her dream of following her was ridiculous. Shut out of higher education, she was stuck on an endless hamster wheel of low-paid, mind-numbing jobs. But she was determined that wouldn't be it.

'Michael Jordan wouldn't be Michael Jordan if he hadn't played basketball,' she explained. She just needed to discover her *métier*.

One day while browsing online for jobs, she stumbled across an advert for a coding course called Web@cadémie. It was a two-year programme for 18- to 25-year-olds at Epitech, France's premier information technology institute. Her hopes weren't high. Without a *bac* and with no money for fees, she'd been shut out of countless opportunities for further education. But this one claimed to be free. Moreover, it was specifically *for* dropouts. The entry criteria clearly stipulated that no qualifications were permitted. Lilas quickly completed the online application and was invited to attend *la piscine*: the Pool. It was a mysterious, intensive three-week non-stop coding marathon that served as the final selection for the course. She jumped in.

The Learners Shall Inherit the Jobs

As natural born learners, it should be no surprise that what we learn, or fail to, during our schooldays shapes the trajectory of our lives. For every dropout-done-good like Richard Branson or Steve Jobs, there are a million kids who didn't make it, struggling to get by. Mastering the basics of English and maths like Ifrah is necessary, but it's no longer sufficient. In the next stage of our journey, we'll explore what else we should be learning. Experts forecast that by 2022 the UK will be home to '9 million low-skilled people chasing just 4 million jobs', and experience 'a shortfall of 3 million workers for the higher-skilled jobs'.[1] In the last two decades in the US, 7 million unskilled jobs have been shed from the economy, with openings for unqualified workers down 50 per cent in ten years.[2] Frey and Osborne, the Oxford Martin School boffins we met in chapter 1, estimated that half the jobs in Western economies were susceptible to automation in the next 40 years, along with *three-quarters* of those in China and India.[3]

It seems like now might be a good time to panic. But we shouldn't succumb to the inevitable just yet.

The seers of the nineteenth century were as alarmed by industrialisation as we are about digitisation. The economist David Ricardo warned that 'the substitution of machinery for human labour' would be 'very injurious to the interests of the class of labourers' and 'may render the population redundant'. But work never did disappear. New jobs were created that demanded learning. 'Children of displaced handloom weavers not only had the option to work in machine-intensive cotton mills,' wrote historians at Northwestern University, 'they could also become trained engineers and telegraphy operators.'[4] The new roles contributed to a nineteenth-century learning explosion and the growth of universal schooling. We adjusted our approach to the human-capital needs of the new economy, found things to do.

Our challenge today is greater. There are many more people on earth for a start. But new roles do spring up. Half the jobs in the best-paid quartile of vacancies in Western economies are now 'hybrid' positions, requiring frequent use of digital skills.[5] Marketing professionals and graphic designers are expected to program algorithms. Demand for data analysts is up over 300 per cent and for data-visualisers (unlikely to have featured prominently at your school careers day) a whopping 2,500 per cent. In 2016, when the German engineering firm Siemens opened a plant in North Carolina, 10,000 people showed up at a job fair for 800 positions.[6] There *are* new roles. You just have to be smarter than a robot to get them. Sadly for the candidates in this case, only one in seven passed a test of 14-year-old-level maths, reading and writing. Better luck next time.

Part II of this book urges us to *do* better. It's impossible to predict the jobs of the future, or whether we'll even work. But it's perfectly possible to see that we can better prepare kids for the world to come. Today we accept that some of them depart school at 18 with *nothing*. No qualifications. No knowledge of the world. No skills. No direction at all. *So long and thanks for the memories!* We limit their opportunity to learn to a brief period at the start of life, often pushing them off into the world with a sense that they've already failed. I hope to show how that could change if we place the ideal of *craft* at the centre of our learning efforts. It means helping every child to find their purpose, express themselves creatively and master the tools they need to do so. Our journey starts with the idea of lifelong learning and the story of Lilas Merbouche.

To find out what we could do differently to set our kids up to succeed in the high-tech roles of tomorrow, I hopped on a Eurostar from St Pancras to visit a revolutionary new institution in Paris that was doing just that.

So Long, and Thanks for All the Fish

'The system is broken,' read a charred panel in a stairwell off the main lobby as knots of students – the about-to-smoke, the had-just-smoked – passed by in scuffed Doc Martens with their hoodies up and backpacks on. 'Obey!' commanded a monochrome image of André the Giant from the far wall. On the steps of the metal-grille-clad building the smokers were always visible, *clops* burning day and night in an endless relay. The school never slept. Inside on the wall a stencilled security camera asked, 'What are you looking at?' as a 30-something in a green sweater washed down two Burger King value meals with a bottle of San Pellegrino. The palette was cyber-punk, the place was 42. Named after the 'Answer to the Question of Life, the Universe and Everything' in Douglas Adams's *The Hitchhiker's Guide to the Galaxy*, it was a new university out on the windswept Boulevard Bessières in northern Paris and a temple to our new religion of coding.[7]

Thomas, 23, met me in the lobby. A second-year student and aspiring tech entrepreneur in blue jeans and a grey sweater, he went by 'greem' in his online gaming circles, after Grimbold, a minor hero in *Lord of the Rings*.[8] He'd got into coding after spending his teenage years in La Vendée repairing computers in his parents' garage. At 42, he was living his dream. One of the building's three floors was 'Middle Earth'. The others were 'Westeros' and 'Tatooine'.[9]

'The brain and the heart,' he announced, gesturing mock-grandly to a black box within a huge sealed glass cube.

Green and red LEDs blipped on and off. 'If that stops, the whole thing grinds to a halt,' he went on. *That* was the server. It housed the 'Intra' whose algorithms powered 42. The school's software was continually analysing student performance data from the projects they worked on all day and, frequently, all night. It was self-organising, using the information it gathered to make constant, automatic micro-adjustments to the student-user experience. And it had developed a caring side. Noting a recent positive bump in the results of students who'd taken meditation classes, it was now subtly suggesting mindfulness to all. I thought it might tackle their nicotine habit. The Intra was 42's Skynet. It set student projects, ensured they were keeping up, prompted them to collaborate, as they would in the world of work. I just hoped it wouldn't rise up and try to take over the world.

That wasn't as far-fetched as it sounded. The Intra also facilitated 42's boldest selling point. The school had no teachers.

42 had launched three years earlier with a radical vision. Asserting its freedom from the constraints of French education, it would have no teachers, no fees and no prior qualifications required for entry. Now home to 3,000 mostly male, not all geeky students of 18–65 (30 places were held for over-55s who'd been long-term unemployed), it promised an intensive, creative, self-guided learning experience. You left at the end without a certificate but with web development skills desperately needed in the modern workplace. When the school opened a few years earlier, there were 60,000 coding vacancies in France.[10] In 2016, 30,000 hopefuls had taken the online logic tests that constituted part one of the application process for 1,000 available places. All of 42's graduates were employed in well-paid jobs in the tech sector.

'Cool, eh,' said Thomas. He pointed to a rugby ball on top of the server. It had been signed by the French forward Sebastian 'The Caveman' Chabal. But the man-mountain was small fry compared to the list of 42's other fans, which read like Mark Zuckerberg's LinkedIn profile. Elon Musk had videoed in. Snapchat's Evan Spiegel couldn't stop talking about it. The CEOs of Twitter, Periscope, Airbnb and Slack had all lavished praise on the place. It was *unicorn*-hot. As we walked out past the server and into a vast bright room dotted with Banksys and lined with rows of gleaming white desks, I found myself seduced by the futuristic vision. Hundreds of pristine iMacs hummed in quiet unison on the desktops. Around the terminals students sat in ones and twos, headphones on, eyes glued to the screens. They were buzzing. With no lectures to be late for, you worked the hours you wanted. 8 a.m. till 6 p.m. Midday to midnight. Whatever. It looked, as the school's website advertised, like they were *born to code*.

So, it turned out, was Lilas Merbouche. The students didn't know it, but they owed their education to a dropout shop assistant from the banlieue.

'I was there from 8 a.m. to midnight every day for three weeks,' Lilas laughed when we caught up via Skype. Now 27, with long brown-blonde hair, a no-bullshit manner and TV presenter smile, she radiated contentment. Five years after making her application, she was now *running* Web@cadémie. Lilas had found her element. 'I loved what I was doing,' she told me. Faced with a slow life in the checkout lane, she entered the *piscine* on a mission. There everyone had started from zero and she threw herself into the daily challenges, working every waking hour and flying through the selection process. Coding was her basketball. She fell in love with the creative side of it. Where school work was abstract, this was tangible. 'You write lines of code and it *results in something*,' she said. For two years Lilas worked 16-hour days, and often all weekend, always with a smile on her face. Graduating top of her class, she got her big break, winning a developer job at Free, France's second-biggest internet service provider and third-largest mobile operator.

Founded by the now billionaire Xavier Niel, Free was as lean as a start-up. Its hundred employees were dwarfed by the 15,000 at Orange France, though both did much the same work. It was also flat – non-hierarchical – and Lilas soon found herself thrown into new-product meetings with the big boss. Niel was taken by her contributions. Lilas saw things differently and wasn't afraid to voice her opinions. When he asked where she'd come from, she told him about the pet shop and the coding academy. 'That's when he called me,' said Nicolas Sadirac, 42's co-founder, when we met later. He had started Web@cadémie and vaguely known Niel for 25 years as a fellow traveller on the worldwide web. 'I don't understand,' Niel had told him on the phone. 'I'm looking for talents everywhere and I don't find any. All my friends are doing the same thing. We don't find any. And there is a *girl*,' he pronounced it flamboyantly in a French accent, 'who is *wonderful*, who is *full* of talent and she ends up selling pets rather than doing information technology. What's going on in this country?'

Sadirac told Niel he should invite him for lunch to learn about his work. 'I ate in a very good restaurant,' he chuckled.

'He asked me how much,' Sadirac said. How much would it cost to set up a free coding academy that would be accessible to every kid in France, no matter their background, provided they had the desire and willingness to learn? 'I told him 100 million euros.' It was a ballpark figure, tossed out over the petits-fours but based on long experience running tech schools across the country. By the time the espressos were served Niel had made up his mind. 'OK, we do it,' he said. He didn't think it was expensive. But he had one proviso, with Lilas in mind. 'Make two girls that way each year and it's worth it.' Niel ultimately invested 70 million euros, enough to see 42 through its first decade. The subsequent plan was for alumni to provide the funds in perpetuity. Many are already contributing.

'The French system is broken,' wrote Niel in an editorial for the school's launch, 'stuck between universities that are free and accessible to all, but not always aligned with the needs of businesses', and private schools, which were 'more effective but also costly, leaving by the roadside a huge amount of talent – genius, even – that might be found in France'.[11] While critics derided universities as 'unemployment factories',[12] failing to prepare kids for the economy of today, 42 would solve the problem, at least for web developers. It recently opened a California campus and had plans to grow worldwide.

Back in the elevator ('*l'acen*-swag,' said Thomas, grinning) disco lights and Daft Punk accompanied us to the first floor, where he logged on, shielding his password. 'Keep Calm and RTFM,' read his desktop background. *Read the Fucking Manual.* In the internet era, the answers were out there. You just had to be bothered to find them. He opened a large map showing two concentric circles on a black background, with a central spot from which branched out sets of connectors, nodes, sub-connectors and sub-nodes. Like something out of *Star Trek*. It had the logic of a computer game. Each node was one of around a hundred projects that could be unlocked only by completing a prior challenge. You started in the middle and worked out. Thomas was 16 per cent through level 5. He flashed up a mesmerising animation of fractals that he had been programming – one of the self-guided projects. There were 21 levels in all.

I reflected on what it might mean to have a 42 for every occupation. The lift vibes and mythologies might differ – psychiatrists or nurses might not be *that* taken with the space-elf theme – but the practices felt transferable. You could bring together aspiring novices of every age, set them practical challenges of increasing levels of difficulty, prompt them to evaluate each other's efforts *et voilà*, leave them to it. Without traditional hierarchies to slow things down, including teachers trained on outmoded knowledge, it might be easier to keep pace with the shifting needs of the world. Wasn't a big part of education about trying on risk-free the cloak of a chosen role? Lawyer-students wore suits. Sociologists grew dreadlocks. You'd still get all of that, only you'd leave with real-world skills developed through the projects. Perhaps it wouldn't replace current provision for schoolkids, but it could add to it throughout our lives.

We were heading for 'Valhalla', the lone tech-free zone in the building. I was finding it hard to get my bearings. It was so *new*. But a part of me hesitated. We couldn't all be coders. We didn't all want to be. Didn't some of us have to be basketball players? Lilas had found her *métier* because she'd eventually had a taste of something she'd liked, the activity that gave her purpose. Most kids never got that far.

As Thomas opened the door into a room of soft greys and

mood-lighting, I tried to imagine how we could give them a better chance of doing so. I wasn't the first to have that thought.

The City Where the Kids aren't Exactly in Charge

On a pleasant Sunday not long ago I met with my wife's cousins Jacob, 14, Sofia, 11, and Tor, four, to head deep into Westfield shopping centre in London's White City. An airy, anodyne vision of Dubai Airport urbanism – but without the gold-plating – the sprawling development boasted the usual clones (Apple Store, Top Shop, Starbucks) along with a high-end retail zone (Louis Vuitton, Burberry) called the Village. All your wants met, all under one roof. But the three siblings, who had driven down to meet me with their parents, skipped by the Lego and Uniqlo stores without a second look. They were headed to KidZania, a *child-size city!* where through *sixty role-play adventures!* they would *work, earn and save!*

At a scaled-down British Airways desk Marie took our boarding passes and briefed the young flyers.

'It's a bit scary,' said Jacob, of the grey and orange bracelet she'd snapped around his wrist. The tamper-proof electronic bands stopped kids escaping the perimeter unaccompanied and, more probably, prevented parents (who wore one too) sidling off to browse for Nikes.

Sofia and Tor excitedly counted the 50 KidZos they'd been handed at check-in. A KidZanian universal benefit, the Monopoly-style money gave each child a starter sum to spend on activities in the land beyond. If they wanted more KidZos, they were expected to earn them through the simulated work experiences that were KidZania's central attraction. If they reached 75 KidZos they could open an account at the Central Bank of KidZania and receive a fully functioning debit card.

Behind us, the fuselage of a jumbo jet dominated the check-in plaza. *To fly. To serve.* It was home to the BA Aviation Academy.

'Kai,' said a uniformed border guard, one of the Zupervisors, placing two fingers over his heart in salutation. Here we go, I thought.

The nation of KidZania was born in Santa Fe, Mexico, in 1999 as La Ciudad de los Niños ('the City of Children'), a venture by two Mexican entrepreneurs and friends from first grade, Luis Javier Laresgoiti and Xavier López Ancona. The origin myth claimed a historic Declaration of Independence – of kids from adults – and inscribed six inalienable 'Rightz of Childhood' guaranteed by 'RightZkeepers' Urbano, Beebop, Chika, Vita and Bache: to be, to know, to create, to share, to care and to play.[13] Though primarily a business, and a successful one (800,000 visitors came in the first year and it now boasted 24 mall-based micro-worlds from Seoul to Mumbai, Chicago to Kuwait), it was this promise of serving kids ('We are empowering them to be independent,' López told the *New Yorker*) that had brought me to the park.[14] KidZania London was run by a director of education, Ger Graus, OBE, who had assembled an advisory group 'think tank' of some of the country's leading educators. His tagline: 'Children can only aspire to what they know exists.'

KidZania claimed to take its role as a terrain of real-world learning seriously, aspiring to open kids' eyes to opportunities they'd never have at school, or in life. Was it working?

We strolled out onto a surprisingly closely realised two-thirds-size simulation of a faux-colonial boulevard along which three pint-size police constables were hurrying. A line of wrought-iron streetlamps led to the supermarket, where five-year-olds were solemnly placing produce into tiny trolleys. Through the window of the hospital's A&E Department to our left, four medical students could be seen removing the liver from a dummy patient. On our right an employment office screen advertised job openings: 6 KidZos for ten minutes' work as a beautician, 10 for bank tellers, 5 for air-conditioning technicians. The sealed 75,000-square-foot space was split over two levels and offered jobs for aspiring artists, smoothie-makers, recyclers, supermodels, dentists, mechanics and journalists, all earning – or costing – a specified amount.

At the university, a two-storey building in the main plaza shaded by a fake tree whose leaves brushed the painted-on clouds, they could earn a degree. It added 2 KidZos to their wages for certain jobs.

A fire engine passed, piloted by a Zupervisor. 'Zaz!' she called out, a KidZanian exclamation. Six recruits in navy-blue coats and yellow hats were preparing to battle the half-hourly recurring blaze at Hotel Flamingo as a pair of bleary-eyed radio journalists emerged from the station into the permanently falling, climate-controlled French Riviera dusk. It *was* a city – a moon colony, even – for kids. They went about their mock-adult business with real seriousness.

'Kids are organised,' Ger Graus said when we'd met a few weeks earlier. Wearing the brown blazer, small circular spectacles and swept-back hair of a children's book wizard, he was exactly as you'd imagine the impresario of a child-size city to be. 'Chaos comes the minute a grown-up steps in.' After a brief tour of the domain, we'd sat down for a coffee in KidZania's only child-free zone, a café on the first floor near the Al-Jazeera Newsroom, where adults gathered to peer soothingly into screens while the kids were at work. In learning terms, Graus was Jean-Jacques Rousseau meets Willy Wonka. He felt passionately that our systems misunderstood kids, pressing them to conform. *Kids are born free, but everywhere are in schools.* 'Trust the children,' was his mantra, 'they trust us.'

Graus felt KidZania gave kids a chance to try things out, broaden their horizons. Growing up in south Holland he'd wanted to be a coalminer like his grandad – 'that's what most of my family were' – just as in some places you expected to be unemployed because all your family were. His other dream was to be Johan Cruyff, the Dutch football maestro. 'I became neither,' he added. Instead, his German teacher persuaded his parents that he should apply to university and become a teacher. He had taught for years in Norfolk and Humberside, before becoming an Ofsted inspector and leading two Education Action Zones in the north of England, raising attainment across 30 schools near Manchester Airport.[15] In 2014 he was awarded an OBE. Such are the paths of our careers.

'My experience was quite flukish,' he said. Shortly after he'd graduated, Holland's coal mines had closed. In another life, he'd have been on the slag heap and he was concerned that we essentially left these things to chance. KidZania aimed to plant more seeds in a child's mind than just footballer, miner, teacher or coder. In another world, schools would create this base from which children form their own opinions.

How *did* we expect kids to know what they wanted to do, if

they experienced so little beyond the GCSE curriculum? With my Year 10s, we'd once staged a CSI experience while studying Sherlock Holmes. It added nothing to the quality of coursework, but the kids had come alive trying on the too-small bodysuits and sifting through the evidence. I began to wonder if beyond all the pageantry somewhere like KidZania did have a role to play.

At the Aviation Academy, Jacob (a head taller and eight years older) opted out of joining the three-foot air cadets. Tor checked in with his electronic bracelet and handed over 10 KidZos. A utopian quirk of KidZania saw the most pleasurable activities (learning to fly, making ice-cream) costing more, and the least (recycling, stocking shelves) paying most. In the parallel activity, flight attendants earned 8 KidZos for looking after passenger-parents. The split was predictably gendered.

'Are you excited to be pilots today?' called out Flight Instructor Nadia. Zupervisors, played by chipper students and struggling actors, delivered scripted content prepared by experts on the education team from guidelines given by the corporate sponsors. The kids loved her.

'*Yeah!*' screamed the seven tiny pilots. They'd been standing to attention in their ill-fitting pilot's uniforms, but now broke into applause.

Nadia continued with some key knowledge. The simulators were real, the same ones used by actual BA pilots. They clocked up *3,000* hours and *two years* of training before they could fly a real plane. The A380 needs 2,300 litres of paint in *five coats*. That weighs the same as *three gorillas*. The budding airmen crowded round as she brought her plane down onto the runway. She asked some questions. Who wants to be a pilot? What is this switch for? Do you need wheels in the air? The pace was high, in keeping with the 20-minute window in which most activities were completed, but I thought the experience was a little underwhelming, a bit too rigid, rewarding conformity over experimentation. Great for certain jobs, but less so if you aspired to find your purpose.

Tor inexpertly brought the plane in to land, flicking at the overhead switches as he taxied off the end of the runway into the long grass.

'I *love* flying,' chirruped a pilot at another of the five consoles.

'Unbelievable flying,' added another.

'That was *cool*,' said a third.

Nadia assembled them in *one straight line* before their exuberance got the better of them, handing each qualified pilot a set of cardboard wings. They all beamed, but Tor was a little nonplussed.

'It was good,' he said, 'and a bit bad.'

This was exactly the point, stressed Graus. His own daughter had done work experience helping out at a primary school. 'After two weeks, if it was the last job on earth, she wouldn't do it,' he laughed. The school phoned to apologise that the work experience hadn't been a success. 'Nonsense!' he'd replied. 'It's the most successful thing she's done since she's been at your school!' It was the first time she'd been put in a position to decide something for herself. 'We have this odd view of education that you have to pass everything,' he added. In preparing for life, kids needed to experience as many options as possible in order to find what they loved, or didn't. These opportunities for exploration were increasingly rare, especially for poor kids.

I agreed, but it wasn't clear that kids were getting the full experience at KidZania either. At the richly realised Al-Jazeera Newsroom, Sofia's experience had been limited to hitting a couple of buttons while Zupervisor Olly – in the country of the kids, the biggest kid is king – ran the show. It rankled. KidZania *was* a business first of all, but Graus had also told me about a 'Dirt is Good' campaign that he'd advised on with education guru Sir Ken Robinson to encourage kids to play and explore outside.* Children had the run of KidZania, but in the activities, there was little room to be creative. Some experiences afforded a little imagination – at the Pokémon Animation Studio, Jacob created a virtuoso short film of a tree – but others – walk the H&M runway at a fashion show, 'make' Cadbury's at the

---

* Sir Ken Robinson is a wry British academic and darling of TED (his talk on learning is the most watched of all time) who has worked tirelessly to encourage our schools and education systems to embrace a new idea of 'creative learning', arguing for example that we should value dance, say, as highly as history in the development of our kids. I highly recommend watching his talk, then coming back to these pages.

chocolate factory – seemed more of a ploy to brainwash kids into brand loyalty. KidZania had 800 corporate sponsors worldwide, with Renault, Cadbury, Innocent, Pokémon, H&M and British Airways all prominent on the Elysian walkways of KidZania London.

I walked past the Ministry of Sound, where a dozen kids were boogying to 'Shake It Off' by Taylor Swift. I had mixed feelings. What did the kids think?

'It's like a really relaxed school, where you're not being judged,' said Jacob in the quiet of the auditorium. He thought he was a bit too old for it, but was still impressed by the quality of the equipment. The KidZos were a motivator. 'It's a bit like Monopoly,' he added. Sofia had accumulated 80 and opened a bank account. She didn't think it was educational in the sense of 'Wow! I never knew that' and she also thought that KidZania ought to have a bakery ('a real one, not a fake one'), where kids could make cakes. But she liked it. 'It gives kids a chance to experience something they've never done before.' Jacob preferred the animation studio, but her favourite activity had been recycling. Why? 'Because it was short and you get 8 KidZos.' She really was learning some lessons about working life.

Unexpectedly, the most popular activity was the UPS experience. Of the 400,000 kids who visited in the first year, 200,000 had donned the branded caps and high-vis jackets before hitting the imitation tarmac open road. 'If you're an eight-year-old boy you can do it with your mate, you get paid, get a uniform, nobody hurries you, you're in dreamland.' I understood. We were at heart a race of white-van drivers who cherished freedom, wanted to explore.

Graus thought that the first question we ought to ask in schools was 'What do we think by age eleven every child should have an experience of?' It was a valuable one. We needed to get away from a world where we told kids what *not* to do, he argued. 'Don't walk on the left. Don't shout. Don't laugh.' Instead let them experience what they could do, a bit like in Pen Green. The kids *did* have some meaningful experiences in KidZania. Sofia operated on a medical dummy supervised by a real-life qualified doctor. Jacob found his niche in the Pokémon Animation Studio. Tor put out a fire, and got a tattoo. KidZania had given them a taste of the world *out there*. But only a very partial one.

As we were leaving, Sofia went with her Central Bank of KidZania debit card to buy a bumble bee key-ring at the two-thirds size department store. Finding she was a few KidZos short, she stopped and headed for the nearest place to earn more cash, the H&M clothes-recycling experience. She patiently tapped her foot through the scripted intro, dutifully experienced the feel of the materials for the second time, and then began to tap away on a tablet – the work – sorting virtual clothes into piles of wool, cotton or nylon to secure the 10 KidZos she needed.

'Maybe we should get kids doing more of this,' said Mum Eleanor, 'then they'd understand the value of money.'

Across the world millions of kids *were* destined for this work. Without access to a good education, jobs increasingly automated, trapped by circumstance, sorting waste clothes was all they could aspire to. Even with employment, adult life might mean hours of menial labour, contributing to someone else's millions, all to save a little money to buy needless tat from a shop. I thought that told us something valuable about finding our purpose somewhere, even if it wasn't in our work. Otherwise, this was a scripted reality, too free of risk to really give kids a chance to learn, a simulated world nesting within the already simulated world of Westfield. No sky. No grass. You *could* have been on Mars. Jacob, Sofia and Tor had a great time, but they didn't have much of a chance to learn to be, to know, to create, to share, to care or to play. You couldn't begrudge KidZania the concessions to consumerism too much though. We all had to make a dime.

Better in the long run, I thought, if we could bring these experiences into schools. KidZania gave kids a micro-taste of different careers. What would it mean to help them find their place in the world?

The Future Belongs to the Learn-It-Alls

'The learn-it-all will always trump the know-it-all in the long run, even if they start with less innate capability,' said Microsoft's CEO, Satya Nadella, in an interview in 2016. He was borrowing from a Stanford University psychologist, Carol Dweck, whose theory of 'fixed' and 'growth' mindsets was taking off in schools worldwide. Those in the first camp believed like Francis Galton that talent was inherent, while those in the second thought that you could always improve your performance through effort. Dweck asked us to picture two kids in a classroom. The know-it-all begins ahead. She feels smarter than other kids due to her superior knowledge. Her fixed mindset means she believes her advantage is down to an innate superiority and she doesn't work hard, thinking her talent comes naturally. The learn-it-all on the other hand sees talent as a result of effort. She has a growth mindset. She strives to get better, to add new skills, to deepen her knowledge. In the long run, just like Ifrah, she overtakes the know-it-all.[16]

Today it no longer matters what you know, but what you learn. Rapid technological change, ever-longer lives and the robots that are right now coming for our jobs mean that there is a premium on learning itself. Like 42 and even places like Eton College, which also weigh kids' character and ability to contribute to the life of the school in its entry criteria, companies like Google are looking beyond college degrees, using their own assessments to select candidates.[17] According to Google's CEO, Eric Schmidt, they look primarily for 'learning animals'.[18] The UK's Big Four professional services firms – PWC, E&Y, KPMG and Deloitte – are also looking past what candidates have earned in their degrees, just as KSA looks past their Year 7s' current grades. Instead, the firms' selection criteria now put more emphasis on candidates willingness to learn.[19] The learners will inherit the jobs.

The good news is that opportunities for ongoing learning are opening up. Harvard University and MIT partnered to launch the EdX online learning platform, where students of any age can take micro-masters. Though the hype over MOOCs (Massive Open Online Courses) has died down,* they are still a steady source of

---

* In the early 2010s MOOCs were proclaimed the future of learning by a wide range of commentators. They promised free access to the world's best university courses. No longer was learning restricted to being in the lecture room with, say, Michael Sandel at Harvard (you can look at his excellent lectures on 'Justice: What's the Right Thing to Do?' online); you could now watch them from the comfort of

basic online qualifications used by millions of students worldwide to develop particular skills. General Assembly has for decades offered short career courses. You can purchase a spot in an online Master-Class to learn film-making with Werner Herzog, comedy with Steve Martin or cooking with Gordon Ramsey. If you're spiritually bereft, you can find your place in the world at Alain de Botton's School of Life.[20] Lifelong learning is de rigueur and increasingly big business. The learning is out there.

The bad news is that accessing those opportunities is a question of skill and will. You have to *want* to learn, and you need to be able to do so. In our schools, we are too often crushing kids' desire and interest, as Ger had pointed out, or making them miserable, like Lilas. Lifelong learning means being a learn-it-all, but it also means *loving* learning. Everyone I visited worried that schools were teaching kids to hate it. Companies weren't on top of it either. Despite the extra value you could grow through learning, the average British worker was receiving an average of just 40 minutes of training a week. It has to change.

We have two options: teach kids the skills that are needed and risk them falling out of date, or teach them to teach themselves the same skills *as and when they need them*. It isn't about giving kids a fish. Or a fishing rod. We need to help them to develop the ingenuity, adaptability and self-motivation to be able to figure it all out for themselves. Training someone early to do one thing all of their lives is no longer the answer. Instead our kids have to focus on learning how to learn. Doing better means ensuring that all kids had the chance to master the tools they need to achieve their purpose, then helping them to see the deeper principles of learning that underpin that process. That's what had captured my imagination at 42. There the students were in their element.

---

your own living room via the internet. Despite some exciting early successes (when Stanford University opened up one of its computer science courses to allow any student with an internet connection to study alongside its undergraduates, the top 400 papers submitted at the end of the course came from *non*-Stanford University students), the promise of MOOCs has not been realised. The problem is that lots of people sign up, but very few – just 5–10 per cent – ever make it to the end.

All Watched Over by Machines of Loving Grace

In Lothlorien, a room named after the elf kingdom in *Lord of the Rings*, Nicolas Sadirac sat forward in his chair. 'I don't,' he said with a shrug, when I asked about how he ran 42. His wild hair and tanned face framed a pair of piercing blue eyes. 'It's the software.' Despite the cold outside, the atmosphere was humid. A living wall thick with palm, fern and ivy emitted a fine vapour from behind the whiteboard. Resembling an organic wine-maker in his scuffed boots, jeans and 42 T-shirt, Sadirac seemed an unlikely founder of the world's most avant-garde school. But he was a *grand homme* of the tech republic, a cyber-security expert and former leader of the nation's top information technology institutes. He'd even infamously hacked the website of the Prime Minister, Lionel Jospin, at the turn of the millennium. The Intra really did run 42.

'We just try stuff. It works? We put it in place.'

After his PhD in physics at Stanford University, Sadirac had voyaged home via Japan determined to get into the new craze of computing that was sweeping Tokyo and San Francisco. It was the tail-end of the Eighties and in France information technology was an arcane subject taught only in specialist universities. 'It was very theoretical, without *any* computer,' he said. Sadirac shopped around, choosing EPITA, the Advanced School of Information and Technology, as the only one that actually *had* a working computer. But it was a private school and he couldn't afford it, so he struck a deal. 'I teach maths and physics and I get the computer class.' EPITA agreed. He sought out as much real computer time as possible – the classes were poor – but it was in his teaching of undergraduate stochastics, statistics and probability that he made his most important discovery.

One day Casino – the French supermarket giant – approached EPITA with an optimisation project. How to reduce the time people waited to pay? When the director told Sadirac his class would take it on, he protested. 'The students are dummies! We will be ashamed of the result!' he warned. The students were averaging 2 out of 20 in their tests – 'They were just awful,' he remembered – and the maths required for the project was way beyond their ability. The director insisted. 'So I did it.' He shrugged. Students were put into groups of four, given a brief and set to work. 'And they achieved *good* results!' Sadirac exclaimed, sounding surprised even after 20 years. He wondered what was going on. Had they cheated? To make sure, he gave them an exam. In it, the dunces scored 12s, 15s out of 20 – not *good*, but much better.

It was a revelation. If you gave the students a project, *they worked out the maths part.*

'I didn't teach any more,' he said, 'I didn't do a class in-between, they just did the project.'

Over two decades, Sadirac refined the approach at Epitech, Web@cadémie and later 42, experimenting with methods, teaching configurations and evaluations. Students did best working on projects rather than learning theory, learning from other students rather than teachers and receiving their grades from peers. 'Students grade harder – much harder – than the teachers.' To make it possible, you just needed the right technology. Sadirac felt it was close to Sugata Mitra's idea of a Self-Organising Learning Environment. It was. And, at least for the students who'd made it through the *piscine*, who, granted, were among the most self-motivated, it was *working*.

On the top floor of 42, Thomas had shown me *le bocal* – the Jar – a large glass office stocked with towers of supplies – Red Bull, instant noodles, Coca-Cola, towels, Nerf guns, Lucky Strikes – that suggested its inhabitants didn't intend to leave very often. 'Your Empire Needs You,' read a Darth Vader poster on the wall. This was 42's nerve centre, a microcosm of life in the digital world. The half-dozen developers sloped coding at their desks in black hoodies headed outside to the roof-deck, where a hot tub awaited Friday evening festivities. They could crash overnight in one of eight bunk beds curtained off from view, keeping a constant eye on the Intra. Beyond that it was down to the students. Begin with the map. Decide where to go. Make a team. Do a project. Receive an experience point. Go up a level. A little, I suppose, like Angry Birds. The software kept all the data, and made it accessible to every other student on the block chain. The learning of one student added to the learning of *all*. They could append notes, change that project.

'It's a full self-learn system,' said Sadirac, who sat on a zebra-print throne near a stash of hard liquor and exotic herbal teas.

'You need more and more creative people in IT,' Sadirac had said. We did. In the UK, Google was complaining that it couldn't find local staff for its operations. Coding was a creative medium – a tool with which you could create almost anything. In the past, programmers ran IT support, today they imagined – and realised – our world. 'Just like in the past for scribes,' he said, pronouncing it *screebs*, compared to today's writers. 42's promotional video predicted that the Shakespeare of the twenty-first century would wrangle code as the bard did words. Sadirac likened 42 to the art schools of antiquity. Creativity and peer critique seemed crucial to his vision of learning throughout a lifetime. 42 was about a learning methodology, not developing particular skills.

'We should not do it in school,' he said adamantly. 'Kids will be disgusted with it!' The French education ministry was planning to introduce coding classes for eight-year-olds. Well intentioned perhaps, but he thought the effect would be to kill kids' passion for it. 'We will break some talents much earlier. I think it's *dangerous*.' It would be like teaching art and telling kids, *OK, you should put blue there, and red next*, with rules, ten pictures a week. It sounded a little like KidZania. Like that, he thought, 'you won't have any more artists very soon'. I wondered how many teachers were true lifelong learners, evolving their knowledge of the world, refining their skills, how many had been forced by the system to put their colours in certain places? Our schools could be resistant to learning.

'In the traditional education system, one big part of the system was done to discipline people, and to remove creativity,' he added.

We once had use for massive workforces that would be willing to suck up daily misery and not break ranks. 'You have a population disciplined and able to work in factories in totally inhuman stuff.' That had made sense when countries were competing industrially, seeking ever-greater efficiencies from their human workers. But now 'all the bad work is done by computers and more and more by robots'. For Sadirac, it was a reason to shift our focus towards creative value, towards craft. Kids were perceptive. Sure, they might be worse at reading and writing than they used to be. But they were *a lot* better with new technologies. 'They make much better robots than they did ten years ago,' he said, impressed. Some had already reached a level he knew he never would. 'My brain was not trained correctly,' he said wistfully.

It seemed what counted was fuelling a desire to learn in all kids – in everyone, really – and preventing the system from killing it. After that, you just needed to make the opportunities available.

Outside among the smokers, I'd met another Nicolas, a 28-year-old student. Pale-skinned, with no-style hair, scraggly beard and glasses – the caricature of a computer programmer – he told me that he'd applied to 42 after seeing it on a TV show. 'It's totally different,' he said. Following school – 'the French education system is not that great' – he'd ended up working as an attendant at Futuroscope, a theme park near Poitiers famed for its 3D and 4D cinemas. At 42 he was learning the basic programming of a dating site, for which he was building out the search criteria – age, sex, sexual orientation – as well as coding a function that used geo-location to ensure you only matched people within a ten-kilometre radius of each other.

'Now, I *want* to find a solution to a problem,' he went on. The work was hard – especially the *piscine*, where he was 'often tired, often ill' – but it was motivating. You chose what you worked on and the experience points kept you moving forward. There was a simplicity and freedom to the method that fostered responsibility. 'If you don't come you don't succeed.' He shrugged. Important too that it was *free*. 42's new Silicon Valley campus would offer dorm places at no cost for those most in need. The promotional video splashed the madness of a student loan system which had seen a 440 per cent increase in student loans in 25 years with over a trillion dollars lent and a current nationwide student debt totalling 6.6 per cent of US GDP. Now that he was a little older – and had known the quiet desperation of an unloved job – he was prepared to seize it. He was excited about his new career.

'It works,' Sadirac had agreed – and it would work for other fields. After the Casino project, he experimented further. His Epitech students had a communication problem. Businesses had told him the coders were 'wonderful in technology, but when they ask us something, I don't understand'. They couldn't write. 'Mainly they are geeks,' he explained. In school they learned that 'whatever I write I never get a good grade'. Sadirac created several sets of picture cards which could be ordered to make stories. 'A cat comes into a kitchen breaks a glass and then gets out.' He laid out a series of five card narratives, which a set of students had to write out as five one-sentence stories. Their writing was then given to another set of students, who had to reassemble the pictures in the right order.

'They just yell at each other!' he laughed. Just 40 per cent recreated the stories correctly. The kids were *mad*, unable to believe their friends couldn't understand them. It inspired them. 'Put a person in front of some kind of goal which has meaning to him, he *wants* to do it.' Sadirac gave them more writing cards and after six months (the blink of an eye in classroom time) 100 per cent of the students were able to write to communicate their meaning. 'Not wonderful writing, but at least the story.' That was a lot, I thought, reflecting on the painstaking attempts we'd made in my English classroom. And he was convinced it would work for French, business, the creative arts. I'd seen it elsewhere. Peer 2 Peer University had groups of adult learners taking free MOOC courses and meeting in public libraries in self-organised seminars. My brother spent a year at School of the Damned, a self-organised MA course where 20 art students gathered in studios, galleries and pubs to design and carry out their own learning, inviting academics and eminent artists to give their seminars. There they'd gone a step further, creating the aims and objectives for their own course. They had no Intra. No money changed hands. Yet they made a lot of new art.

For Sadirac, the formula was simple. 'Get a goal which is harder and harder. Make a team of students. Make a way of evaluating the stuff which is the most objective possible' (this was harder in art or sociology), then let the Intra run the learning, have the students help each other, and let them find the information they need through Google. I wondered what it might take to have a 42 in every community: for data-visualisers, high-tech factory workers, psychiatrists or yoga teachers. In Singapore, every citizen was now given lifelong learning credits to spend throughout their careers. The SkillsFuture Credit was launched in 2015 with $500 provided to help over-25s upgrade themselves by enrolling in government-approved courses like workplace literacy and numeracy, nursing or cooking.[21] Knowledge transferred through institutions couldn't fail to become outdated. At 42, students and alumni adjusted projects on the block chain, or suggested new ones. The curriculum was in continual update, nimbly avoiding obsolescence.

Before I departed, I caught up with another student at 42, Jean-Luc Wingert. In the years prior to making it through *la piscine* and into the school, he'd made his name as an entrepreneur and writer. His book *Le Syndrome de Marie-Antoinette* was a solemn warning to those who failed to adapt quickly enough to the changing times around them. In the best-case scenario, elites were bypassed. In the worst, they lost their heads. Hierarchies usually moved too slowly to keep up with the world. They were inefficient, fragile and concerned with maintaining the status quo. 42 was not. Sadirac had taken a guillotine to traditional models of learning and arrived at something new. *Vive la Révolution!*

Failing to Learn is Learning to Fail

Our schools aren't clear on what kids should be learning today. Reading, maths and science are non-negotiables, furnishing us with our most crucial thinking and communication tools. Knowledge matters too. We need common points of cultural reference and have to lay down our cognitive architecture. Beyond that, it's a crapshoot. Everyone says we have to *prepare for a future that no one can safely predict*, as if any previous generation has been able to see what was coming. And maybe that is just fine. Education is still a solid bet: UK graduates make £225,000 more over the course of their careers than school leavers with no qualifications.[22] Each extra year of education means a 10 per cent increase in your average hourly wage.[23] *You learn, you earn.* Outside of that we risk getting swept up in a tech tsunami when the evidence points in a different direction.

In *The Future of Employment,* Osborne and Frey showed that in the future the jobs we'd do would focus instead on human development. Yes, the job they rated #2 least likely to be automated was 'First-line Supervisors of Mechanics, Installers, Repairers', but beyond that the only other tech-related job in the top 50 was #32, 'Computer Systems Analyst'. The remaining 48 jobs least susceptible to automation included all of the most human professions.

'Recreational Therapist' came in at #1, with 'Mental Health and Substance Abuse Social Workers' at #4, 'Occupational Therapist' at #6, 'Healthcare Social Workers' at #8 and 'Psychologists, All Other' at #15. Jobs like #13 'Choreographers' or #26 'Fabric and Apparel Pattern Makers', #34 'Curators' showed room for creativity, while #11 'Dieticians and Nutritionists' and #15 'Physicians and Surgeons' were just two of twelve top 50 roles related to physical health. Most promisingly, the list was filled out with teaching roles, including #16 'Instructional Coordinators', #20 'Elementary School Teachers', #37 'Preschool Teachers', and #41 'Secondary School Teachers'.

The continued existence of these jobs doesn't mean any of them will be well paid or easy. Tech promises to dominate the economy. A few trillionaire entrepreneurs – it can't be far off – might still own all the profits. But it does illuminate what might come to be our most meaningful work. Instead of teaching kids a set of skills to secure a job, we have to teach them how to keep on learning and growing as people throughout their lives. It is about more than work. Learning throughout your life means finding your purpose and mastering the tools you might use to achieve it. You aspire to develop and refine your expertise, whether as a coder or a tree surgeon, a nurse or a teacher, a cook or a model-maker. Working with humans or tools, our highest aspiration is mastery of a *craft*. That's what Sadirac had achieved and what Lilas hoped to. In the future we should prioritise *learning* itself as the highest end of education, laud those jobs where we work with people and seek to get the best out of them.

Doing better ultimately means embracing creative learning. In the next leg of my journey I planned to visit some of the places around the world where I'd heard they were striving above all to grow creativity in our kids. I'd begin on the East Coast of the US at the MIT Media Lab, a hotbed of technological creativity, before journeying on to Finland to explore the craft of human development.

CHAPTER 5

Creation

Meet Your Makers

> Every child is an artist. The problem is how
> to remain one when we grow up.
>
> Pablo Picasso[1]

Kids in the Wild

Beyond the glass partition, an infant native of the suburbs of Cambridge, Massachusetts, was conscientiously laying out long strings of blue beads on a wooden table, lost in her work. 'We ask our observers to act as Jane Goodall does while observing chimps,' read the email from Castle O'Neill, Head of School at Wild Rose Montessori, where teachers aspired for 'the children to be working as if I did not exist'. Their habitat was a red-brick store front on Massachusetts Avenue, formerly a massage parlour. Castle's office had the snug warmth of a treatment room. 'The idea is that we never want to interrupt this three-hour work time,' she said. I was to be watchful, interacting with the young humans only if they initiated an exchange.

A girl and boy ran past, clutching home-made paper puppets. A ceramicist, the artist-in-residence, painted hand-crafted bowls, watched by a blonde-haired boy in a blue-and-white-striped top. I sat silently in a small wooden chair, eavesdropping on a troop of four eight-year-old boys.

'You mean sophisticated,' said one.

'Sophista-cookies,' added a second, grinning. He had a punk's fin and ear stud.

'You're acting like green beans.'

As well as talking like beat poets, they were playing with a set of wooden boxes. Each contained an array of black and white tiles mimicking pixels on a screen at different levels of definition. They were rearranging them, learning 'computational thinking'. It was a new 'fluency' the school was testing – a twenty-first-century addition to the Montessori Method. Teachers hovered, occasionally shimmering into resolution to give a guiding hand or offer a comment.

While he worked, Zak was crafting a story.

'Dancing girls,' he began. '*Exotic* dancing girls. Cats jumping through rings of fire!'

'It's called a circus,' said his friend.

'No, it's not a circus,' replied Zak matter-of-factly. 'It's the greatest game ever made on the planet! Dancing girls. Dancing girls *who explode!* Exotic animals and dancing girls!'

He manipulated the black and white tiles into a symmetrical pattern, searching for his next line.

'Exotic animals who explode!'

Around the room, 15 children were engaged in these reveries. The puppeteers had secretly stowed their papers under a table to seek help with scriptwriting. A pair of older girls created a Valentine's Day poster. On the sofa, one boy read *The Complete Book of Dragons* to another, using silly voices. The environment was calm. Pot plants abounded, artworks hung on the walls. There were wooden learning materials and richly illustrated books as prescribed by Maria Montessori. The fabled educationalist initially developed her approach teaching children with learning differences in Rome, before founding the Casa dei Bambini in 1907. A physician and psychologist, she took a scientific approach to child development, observing that children often naturally spent long periods deeply attending to items of interest. She prescribed mixed-age classrooms, free choice of activities for students, uninterrupted work time, no tests, specialised materials, freedom of movement and a 'discovery' model intended to give kids freedom within limits. The idea was to build on a child's natural psychological and social development.[2]

Montessori schools had existed on the fringes of learning for a hundred years, but their reputation was growing as a hotbed of creativity. The *Wall Street Journal* had written of a Montessori mafia in the creative elite, with alumni including Beyoncé, Google's Sergei Brin and Larry Page, Wikipedia's Jimmy Page, and Jeff Bezos from Amazon. In the school, kids were learning to *create*.

Wild Rose was one of seven Wildflower Montessori schools trying to bring the method up to date for the twenty-first century. The dream was for a one-room shop-front school in every neighbourhood in the country, independent coffee shop style. Wildflower had been founded by the computer scientist and entrepreneur Sep Kamvar, who was also an artist, using technology as his media. In 'We Feel Fine', he'd created a program that would forever search the internet for the phrases 'I feel' and 'I am feeling' and collect what people had written afterwards. It had so far gathered 12 million of them. A second piece, 'I Want You To Want Me', scoured dating sites for certain phrases, and randomly displayed them in real time on a large screen inside pink and blue balloons, which floated and bumped into one another. Commissioned for Valentine's Day 2008 by MOMA in New York, it had put Kamvar on the map.[3] He now led the Social Computing Group at the MIT Media Lab, a modern Mecca to creative learning. Wildflower had been its major project.

The punk-boy walked over to the girls.

'Roses are red,' he began. 'Violets are blue. Why it smells really, really, really, really, really, really stinky.'

It needed work.

'You're right Hector, it is stinky,' said one of the girls archly. They walked off.

'Two lessons in a row!' exclaimed Zak. He was now peering through a microscope learning about cells. A teacher sat by his side. 'How many do I have to be in!'

I'd asked Castle what the children studied here at Wildflower.

'Whatever they want,' she replied.

Was this *it*? I thought. Could this be the future of learning? It was hard to believe. Kids could only come if their families could afford annual fees of tens of thousands of dollars. It seemed like a luxury a child could afford if their parents were the kinds of folks who lived in the affluent suburbs of Cambridge, MA. But could it ever be for everyone? Wildflowers were hipster schools. Yet Montessori's methods were proven over a century. She herself had worked miracles with some of Rome's poorest kids and her method seemed tailor-made to cultivate precisely those skills that experts said we wanted in the world today. Kids were collaborating. They were creating. They were communicating in complex, even nonsensical, ways. Isn't this what we were after?

MIT Media Lab was just up the road and I had an appointment there with Sep's team later. Kamvar himself was famously hard to reach, usually off surfing somewhere or making new artworks in his seashore cabin. As a group of artists who'd led a creative team to open a school where kids could do whatever they want, I thought they'd help me to better understand creativity and how we might grow it in our kids.

Observing the students in their natural environment, I was even more excited to talk. This was *different*.

Turning back to the room, I noticed the four oldest boys were crowded around a much smaller kid in a paper hat. He cradled a cardboard box, with panels cut out of three sides. Tired of doing maths problems, he'd made a machine that would solve them for him, similar to John Searle's Chinese Room Experiment.* You placed a slip of paper with the problem written on it in one end, waited a while, and then a new slip emerged with the solution at the other. Magic. The trick was that he was the one working out the problems, the ghost in his own machine. Hence the third panel cut out of the back of the device where he inserted his hand and scrawled the answers.[4]

I chuckled. Nice idea. Just like a . . . hold on. Had he just invented the computer? He was five years old. Perhaps this *was* it.

---

* In this famous thought experiment, the philosopher John Searle imagines a man sitting in a room with an English–Chinese dictionary. A helper passes commands in Chinese under the door, which he then translates using the dictionary and passes back out. Would we imagine the man in the room to understand Chinese? Searle used this to argue that AI (at least so-called Weak AI) cannot be considered to have real intelligence. It only *simulates* intelligence.

Practice Makes Perfect, But It Doesn't Make New

Wildflower's freewheelin' style sat a little uneasily with my hard-bitten outlook, forged in mark schemes and lesson objectives at Walworth Academy. But I was prepared to suspend my disbelief. I'd gone to Boston because I wanted to know how to grow creativity in our kids. Experts on all sides, from the Ken Robinsons who valued it for its own sake, to Brynjolfsson and McAfee, authors of *The Second Machine Age*, who thought it kept us ahead of the robot overlords, held it to be the single most important attribute to develop in our kids today. And though I'd initially been sceptical about Wildflower's unstructured environment, the research increasingly backed a hands-off approach.

Three decades ago, an educational psychologist called Benjamin Bloom began a study of creative talent. The bespectacled researcher identified 120 outstanding young men and women in fields like piano playing, sculpture, swimming, tennis, mathematics and neurology, and set out to investigate how they had learned to be world-class performers. For decades he had urged schools to embrace mastery learning, an approach built on an understanding that all kids could excel at anything, given sufficient time and the right support. His study seemed to support his hunch. The genius of the young protégés resulted not from some divinely ordained aptitude, but from long hours of practice and close parental support.[5]

Thanks to Malcolm Gladwell's *Outliers*, psychologist Anders Ericsson's idea of deliberate practice (which we'll come back to) is familiar to us.[6] But Bloom's study also revealed an unsung but no less crucial factor in virtuosity. Prior to investing their 10,000 hours mastering the structures and disciplines of their chosen field, there had been a starter stage, a 'romance' with their field of expertise characterised by play, discovery and experimentation. They'd had *fun*. 'The idea was to see if I could have a bright child who was well adjusted, getting along with people, having friends – not being single-minded,' said one of the parents he interviewed. The most successful and creative individuals weren't hot-housed from a young age, but encouraged to 'be your own person', 'be well rounded' and 'get the most out of your abilities'. Their parents hadn't *tried* to raise geniuses.

As another psychologist, Adam Grant, put it recently in his book *Originals*, 'Practice makes perfect, but it doesn't make new.'[7] He referenced a study of the families of highly creative adolescents in the US. Surprisingly, kids rated among the top 5 per cent most creative in their school systems didn't come from families that were judged more creative than average, but from homes that encouraged independence.[8] Whereas the typical parents surveyed gave their teenagers an average of six rules to follow, those of the most creative kids gave them only one. Creativity began with freedom.

Neuroscience was beginning to map this in our minds, showing that two distinct brain activities are involved in the creative process. The craft and the graft depend on *convergent* thinking, a focused process of thought requiring lots of controlled attention. But there *is*, as Archimedes found, also a clearly observable pattern of activity in the brain prior to a moment of creative insight – and it does often occur in the bath when we're feeling particularly relaxed.[9] This is *divergent* thinking. Using brain imaging, a neuroscientist, John Kounios, has recorded these cerebral eruptions, observing that they are accompanied by the momentary suspension of visual processing in the right hemisphere and characterised by traces of activity throughout the whole brain. You can only achieve this type of inspiration when you're *not* focused. It depends on associative thinking, meaning our creative capacity is broadly dependent not on depth of knowledge, but on breadth, making it hard to game.

In a piece for the *New York Times,* Grant quoted Einstein's adage that 'the greatest scientists are artists as well'.[10] New ideas come from making connections between fields. That requires space and time to mull. In comparison to other scientists, explained Grant, Nobel Prize-winners, who are known for special creativity or insight in their field, are frequently polymaths. They are 22 times as likely to act or perform as other scientists, 12 times as likely to be writers, and twice as likely to play instruments or compose music.[11] Einstein attributed his greatest insight of all to his associative abilities. 'The theory of relativity occurred to me by intuition,' he told Shinichi Suzuki, inventor of the famous 'mother tongue' method of learning an instrument, 'and music is the driving force behind this intuition.'[12]

'Hear that, Tiger Moms and Lombardi Dads,' wrote Grant. 'You can't program a child to become creative. Try to engineer a certain kind of success, and the best you'll get is an ambitious robot. If you want your children to bring original ideas into the world, you need to let them pursue their passions, not yours.'[13] This was counter to the way we'd come to think about school in the UK, US or Asia. We constantly drove kids on towards their exams, took away their time to play. Something had to change and Einstein agreed. Though his mind was crammed with many years of learning and information, he felt it was the times when he daydreamed, letting his mind wander into new and original associations, that were the source of his major insights. 'Imagination,' he said, 'is more important than knowledge.' How did you grow that?

It was with that question in mind that I had travelled to Finland. Fabled for its progressive politics, open-mindedness and social cohesion, the country was also one of the best places in the world to go to school. And where other countries were battery-farming their kids, the Finns had always been free range.

Finnish Lessons

'Oh, they have their artistic differences,' Mervi Kumpulainen assured me on a wet morning in Vantaa, near Helsinki Airport. Five ten-year-old Tarantinos were shooting a movie on the climbing frame, a hackneyed meditation on cinematic violence. With their puffas zipped and hoods up, they were unlikely *auteurs* – but seemed at great ease with their creative process.

Mervi, the hippest and most laid-back primary school teacher I'd ever met, kept an eye on them from a distance. She put it down to giving them space. At the distant tree line, more children appeared, brandishing branches. 'We're always asking, what is the right level of freedom?' she said. The question went to the heart of Hiidenkiven Peruskoulu, the school I was visiting. It was reckoned among the most innovative in the world. *The Economist* had visited recently, looking for the same thing I was, a glimpse into the fabled Finnish Way.

After its teenagers shocked everyone, Finns included, by proving themselves the smartest in the world in the 2000 PISA tests, Finland had gradually become the number one brand in global learning. Though the country's scores had dipped slightly since then (Finland ranked fifth in the most recent PISA tests, with some commentators fearing schools might slide to the same fate as Nokia), it remained a powerhouse, ranking first in 2016 on the World Economic Forum's Human Capital Index, which judged how effective a country was at helping all of its citizens fulfil their potential, both as workers and as people. Fittingly for a country famed for its fierce individuality (home to Moomins, Marimekko and death-metal *Eurovision* winners Lordi, Finland remains the only place in the world I've seen Eighties-style punks stroll the streets), it had achieved this with an approach to learning that was all its own. Where East Asian education powers drilled-and-killed it, in Finland less was more. Kids started school aged seven. Up to the age of 16, when they attended all-through schools like Hiidenkiven, they arrived at 9 a.m. and departed at 2 p.m. Homework expectations typically ranged from low to none. It seemed the opposite of Rocketship or KSA, models that had been so effective in boosting kids' grades.

Less input, better output seemed to make no sense. Until, that is, you thought about the nature of learning in its fullest sense, and of learning to be creative in particular.

'I'm a recovering TV producer,' joked Saku Tuominen from behind the white oval table in his minimalist office overlooking the Bay of Finland. Through the window a large expanse of grass sloped gently down to the Baltic, its low granite islands dotted with fishermen's huts. The writer of ten books, seven on creativity and the human mind, three on Italian cuisine, and a former television executive of global standing ('current affairs, entertainment, quiz shows'), Tuominen is Finnish creativity personified. As he quickly skirted over received wisdoms ('The world is changing and so on . . . Blah, blah, blah, blah, school is crap') and coined new ones ('the term lifelong learning is wrong . . . it is only the means to be active in lifelong *doing*'), he resembled, in his jeans and grey cashmere sweater, a twenty-first-century spin doctor from a Scandinavian political drama, unwinding at home at the weekend.

'I'm running this *mess*.' He gestured around him, before

backtracking. 'It's not a mess. It's clear. It's relevant.' *It* was Hun-drED, a project he'd launched to celebrate Finland's hundredth birthday in 2017. It would find the hundred leading innovations in Finnish education, gifting them to the country, and the world. He agreed with William Gibson that the 'future is already here, but it's unevenly distributed', and he'd seen the opportunity, with the powerful Finnish brand, 'to do great and meaningful things on a global scale'. Nothing was more meaningful to Tuominen than schools. And he felt the biggest gap in our learning was in precisely his area of expertise: creativity. 'Not everything is crap, but there are things that can be so much better,' he went on, 'creativity, thinking skills, globalisation.' He had been running a consultancy teaching creativity to burned-out, worn-down 35-year-olds, when suddenly he realised it would be better to start with kids.

Tuominen saw the future of learning as a triangle that had to be balanced. At the top was general knowledge. 'History, mathematics, whatever.' This was the slow-to-change corpus of Euclidian geom-etry, Newtonian physics and the literary canon. Thinking skills were next. 'You have to be able to question the things that you learn, you have to be able to combine them.' Third came doing skills. 'You have to be able to move from thinking to doing.' With any one piece miss-ing, you failed to prepare kids for the future. No ability to act and they were consigned to unemployment. Unable to think, they were baffled by the complexity of the world. Lacking general knowledge, 'you end up in a situation where everyone is like Donald Trump'. Our current systems focused too strongly on general knowledge, at the expense of thinking and doing. Even Finland's. He was determined to fix it by amping up creativity, which for him meant thinking and doing at the same time.

'Creativity is the will to improve,' he said, finding his groove. 'If you're an artist, you want to improve the way you express pain, or you want to improve the colour blue, or you want to improve the beauty of your melody.' The will to improve meant experiencing and accepting failure. 'Not everything works.' This was something that was catastrophically absent from schools. He'd mentored a high-achieving Finnish girl recently, 'an A plus-plus-plus-plus student', who'd had a nightmare trying to launch a start-up. School had made her a perfectionist, but the world demanded flexibility. The 'education system was all about right answers,' he added. 'Life is not about right answers.' He was worried that the high performance of the system was setting Finns up for a fall.

'There's an old saying, "If it ain't broke, don't fix it." Wrong!' he cried.

This was the Nokia conundrum. While the company was doing well, it failed to adapt. When its business model began to falter, it was already too late. Learning suffered in his eyes from an even bigger problem. 'How do we know what works?' He worried that the world had become obsessed with measurement. As a Finn he was highly sceptical of it. 'We like numbers. How do you know it works? Because 4.2 per cent of whatever . . .' He tailed off. His point was that the most important things, like curiosity, being able to question, creativity, defied categorisation. 'How can you measure these?' he asked. 'What if you question the measurement? What if you question the *school*?' He worried schools in the UK and US would never be able to embrace creative thinking. It defied the authority of teachers as it defied the authority of grading.

We'd also internalised an unhelpful idea that creativity meant only moments of insight.

'There's way too much emphasis on thinking and way too little emphasis on doing.' Instead of seeing creativity as 'think, think, think, think, think', followed by 'do, do, do, do, do', we needed to understand it as a cycle, 'thinking and doing, thinking and doing, thinking and doing'. Schools did often focus narrowly on the mind. I thought back to my own school days, poring silently over books, solving hundreds of maths equations, and my own halting attempts at creativity, thinking long and hard about an artwork or piece of creative writing, then giving up after a first crap draft. If I'd been taught then that the best work took hundreds of drafts, was a process requiring failure after failure, with slight improvements as you progressed through it, I'd have been much better prepared for the world outside school. And I might also have written this book a bit sooner.

'Doing is a great way to think,' said Tuominen, but only if we were willing to push through failure. 'Problems are never bad. They are great. Obstacles are not bad. They are great. For each and every one of those creates us.' Our schools had become streamlined towards perfectionism such that kids were not free to try and fail, except on tests. In the future, kids had to have more freedom, more chance to fail, particularly if we wanted them to *be* creators. If they were to believe they could be, we'd have to let them try. 'The most important skill we should be teaching in school is growth mindset,' said Tuominen, referencing the concept popularised by Carol Dweck. Try, try and try again. Believe you can get better.[14]

Back at Hiidenkiven, Mervi and I walked over to the climbing frame. She'd studied in the UK at Birmingham University before coming back to Finland to teach. With ten applicants for each place on primary school teacher-training programmes, it was a prestigious job. I'd heard that Finnish men and women ranked primary school teacher top of the professions they'd look for in a partner. 'I thought I wasn't going to be good at it. You have to be quite strict,' she said. But it wasn't a strict I recognised. In the wood-chip under the swing, three boys were miming kicking another, as a fifth filmed on the iPad. The scene was pivotal. With their close friend lying critically ill in hospital after a gruesome bike crash, and with no money for surgery, the gang of boys had turned Robin Hood. Their victim was a shopkeeper. They got ready to go again. Eventually, they'd sell tickets to parents to come and watch the final edit.

'Cut!' came a cry from the other side of the playground. A boy in a Beatles wig and an outsized white ice hockey shirt ran screaming past a second child film crew, pursued by a – what? A zombie? 'They're all making horror films,' chuckled Mervi. The group in the forest had been collecting leaves and twigs and stopped to show her. She reminded them of their names. The zombie movie shoot appeared to be going downhill, with the producer in tears. Mervi gave him a brisk hug. Creative freedom was hard. It also sometimes meant making things that were no good – the final cuts were hardly going to be Nordic *noir* standard. The kids would be frustrated, but they were into it and they were creating. 'Films are very motivating,' agreed Mervi. The whole set-up jarred with my experience. I struggled to see a real learning objective and I didn't see any *rigour*. It seemed that the Finnish Way was, well, a bit chaotic. And I was beginning to understand that this was the point.

'We're repeating all the time, it's OK to make mistakes, it's OK to ask stupid questions,' explained Ilppo Kivivuori, Hiidendkivi's vice-principal, over lunch. If Mervi had been an unlikely primary school teacher, Ilppo, in black drainpipe jeans and black shirt, was even less so. A history graduate with a Masters in theoretical studies, he saw the role of the school differently. He didn't talk of pupil achievement, but, instead, of creating the conditions for individuals to understand their identity, and for the school to create a community group. Often Finnish teachers would stay with the same class for six years straight from seven to 13. 'It's supposed to create a safe social environment.' They would know each other so well that they'd have no fear of failure, no self-consciousness. It was important for creativity. 'When you make something different, it's not working all the time.'

At the lunch table, a second teacher joined the conversation. 'It's not developed if you're not given space. First a tiny space. Then a little space. Then more space. And they are kids. They need to have time for play. We don't demand them to sit down and listen for forty-five minutes. It's nearly impossible for any seven-year-old we think here in Finland.' Playfulness was central to their method. It meant having room to breathe. 'They need some ground rules,' agreed Ilppo, 'but rules have to be meaningful, and they have to match with adult life. We respect what's natural in what age.' They respected kids' desire for freedom, giving oxygen to their creativity and curiosity. Play had been squeezed out of schools in the UK and the US, but here it was cherished. It had a larger purpose. Play lay at the root of creativity, which in turn translated into lifelong success. 'In mathematics, there should be an element that you should be able to create equations that have a practical application that you make yourself. It's not just about solving.'

It was one of the reasons why Finland's teenagers did so well internationally. They didn't master such rigorous content as kids in other countries, but the *way* they learned really made a difference. International comparison tests looked into the skills kids needed in the world today, like real complex-problem solving or critical thinking. Finnish kids aced those. They weren't exam-crunching robo-kids, but were set up as three-dimensional characters with a sense of themselves and a place in society. Finns still had a long way to go according to Saku Tuominen, but it felt like their foundation was right for the future.

On the bookshelves in Tuominen's office had been a vast collection of books aligned by their colour. *Glimmer, Blur* and *Think Like Da Vinci.* In Tuominen's own schooldays, his only passion had been ice hockey. One day, while learning a complicated rule of Finnish sentence structure, he'd asked his teacher what was the point. 'It's not important for everyone,' she answered. 'You can either write beautifully, or you can be average. And both are OK. There are different kinds of people. And you have to decide, which one are you? Do you love language, or is it meaningless for you?' Individual freedom and responsibility lay at the heart of Finnish education. So did joy. In their world-leading training, Finnish primary teachers were taught to play the piano and learned how to ice-skate. Yet, in the UK, our whole approach felt increasingly aligned against those.

'You can either love it or not,' Tuominen concluded. He'd fallen out of love with television and found new purpose in schools. His teacher's words had only returned to him after he'd become a writer himself. Just like Lilas, you had to find your spark.

All Schools Should Be Art Schools

Once you found it, you then had to make the most of it. After Bloom debunked the talent myth, the baton was picked up by the psychologist Anders Ericsson. As a teenager he had played chess against a classmate, regularly and easily beating him until one day, quite suddenly, he didn't. Ericsson was fascinated, and a little peeved. Wasn't *he* the better chess player? If they'd been playing as often as one another, how had their aptitudes changed at different rates? The young Swede decided to investigate. He noticed an insight of Francis Galton's (the cousin of Charles Darwin who had done so much to convince people that intelligence was a fixed human trait), that 'eagerness to work, and *power* of working' appeared as crucial to virtuosity as innate ability. Was it true? Ericsson set out to study a group of German violinists, aiming to distinguish what differentiated the experts from the merely good or mediocre.[15]

A single finding made his name, that by age 20 the top performers had under their belts a minimum of 10,000 hours of deliberate practice.

Today these long hours of drilling are a truism of human performance.[16] But as the idea was popularised, our education systems came under pressure to apply the insight not to creative pursuits but to maths, English and science, piling on the time kids spend in these lessons in pursuit of vital exam grades and eliminating the romance stage in which they had freedom to play with their new tools. In schools, our growing understanding of how creativity works has gone hand in hand with a drop in creative pursuits. In the UK, the number of GCSEs taken in arts subjects fell by 8 per cent in 2016.[17] In the seven years to 2014, the number of kids taking music fell from 60,000 to 43,000, art and design from 211,000 to 177,000 and drama from 102,000 to 71,000.[18] Creativity isn't restricted only to arts subjects, but this trend is symptomatic of a broader narrowing of horizons within our schools. Just as creative pursuits are coming to be seen as more important, we are witnessing these subjects drain slowly from our timetables.

Understanding the importance of hard work is both a leap forward *and* a step backward. As Tuominen said, doing was a crucial and often overlooked element of the creative process. We all had it within us to be an expert, were all born creative. Yet 99 per cent of kids were now perspiring, while 1 per cent were inspired. When the artist Bob and Roberta Smith hand-painted a gaudy sign proclaiming 'all schools should be art schools', he was protesting against this status quo. It didn't only mean that there should be more music, art and drama classes. He wanted to see schools in general embracing a different, more creative form of learning. Over in East London, one was doing just that. It promised kids an education to prepare them properly for the twenty-first century, rejecting received wisdom and focusing on doing better. I went to visit.

Perhaps All Schools Should Actually Be Craft Schools

Creativity wasn't something Oli de Botton associated with long days on the campaign trail. There it was all about endurance. The young Labour hopeful had given up on the party's Iron Bum Test after five years. It required bright-eyed hopefuls to sit through buttock-numbing hours of committee meetings until bureaucratic Stockholm Syndrome set in. De Botton had been too much of an idealist. He loved politics and wanted to change the world, but David Miliband's *et tu Brute?* defeat to brother Ed had been a bridge too far. It felt personal.

'I took it badly, that this world I thought was permissive to change, and should be enabling and brilliant . . .' He faltered. 'My experience was, that's not true.'

In the process he'd met Peter Hyman, a former speechwriter to Tony Blair who'd quit No. 10 after a decade in politics to teach history, at first badly, in a North London comprehensive. The two of them shared a belief that schools in the UK were unfit for purpose. 'The idea that in any walk of life you would say, "My judgement of you as a person is for you to sit down in a room for two and a half hours and regurgitate facts in a written exam is utterly, utterly broken as a system,' Hyman told the *Guardian* of his time in the classroom.[19] Their idealism hadn't been extinguished and the pair decided to pour their frustrated intellectual and emotional energy into opening a new school.

'If you're from the Left,' asked de Botton, 'where else are you going to change the world? This is the good work.' In a deliberate challenge to the failures they perceived in the system, theirs would be a school for *today*, designed to close the achievement gap between rich and poor kids, and to develop 'twenty-first-century skills' like critical thinking, freedom of expression, well-being and creativity. They named it School 21. It leaned to the future, but at its heart was the more traditional idea of craftsmanship.

I met de Botton one October morning after taking the Central Line out to East London, the emptied carriage surfacing among the high-rise residential blocks of Stratford, where the school was located. It had been the future once, in the years before the London Olympics. From the platform I glimpsed Zaha Hadid's Pringle-like aquatics centre and the newly re-inhabited athletes' village, populated now by refugees and young professionals, and not yet fully absorbed into the texture of London life. A huge Westfield mall sprawled away from the station, luring happy shoppers. I'd headed the other way, past the bus station, the Shoe Zone and Sports Direct, joining crowds bustling through the old shopping centre. The school was situated in pre-Olympic Stratford, a zone of car chop-shops and raves at the Rex. With 104 languages spoken on its streets, the borough was the most ethnically diverse in the country. This was a truer picture of the world to come, I thought as I pressed the buzzer at the low-slung school building nestled among terraced houses. Rich and poor, old and new, concertinaed together in a few square miles of sprawling city. From this Zone 3 orbit, hundreds of thousands of workers swung into the capital each day to build towers, clean houses, program computers and manage public relations.

'Schools are the last bastions of command and control, aren't they?' asked de Botton as I sat down to talk. 'The last hierarchies left.'

He gestured to his large office. On a whiteboard, four circles of a super-Venn-diagram were labelled Schleicher, Hirsh, Dewey, Freire. A Cambridge classics graduate, de Botton referenced Hegel in passing and unselfconsciously used phrases like *sui generis*.[20]

'You go to KPMG or whatever,' he continued, 'and there are no offices.' Schools were modelling themselves on a corporate structure that the business world had relinquished. It was evidence of a system stuck in the past, wed to outmoded ideas of productivity. Labour under Blair had governed by the same doctrine, their scientific approach to managing the system resulting in a small improvement in GCSE grades, but no growth on international benchmarks, with data suggesting the UK's 16- to 24-year-olds had among the lowest levels of literacy and numeracy in the developed world.[21] De Botton and Hyman had consciously rejected this zeitgeist when designing School 21. The head teacher's office was the exception.

'This idea that you narrow teaching down, narrow the curriculum down to things that you can organise and blocks of time,' he carried on, outraged at how this approach seemed stealthily to be taking over, 'it's not the answer.' Schools were being fêted for finding efficiencies like Five Minute Lesson Plans or Automated Marking, but in his eyes the desire 'to chunk, make manageable, prioritise curriculum knowledge' obviously served the wrong goal, taking the smooth running of the system as a starting point over the development of students progressing through it. Kids were three-dimensional. Learning was hard. One-size-fits-all jobs were drying up.

'It's pessimistic to say there are few remaining jobs left, so we've got to make sure we shove them in there,' he said. A latter-day William Morris, he felt the machinery of the school system ground kids down, alienating them from each other and their work.

'If you believe the school is a microcosm of the world around it, that's not life.' At least it wasn't how he wanted life to be. We had forgotten lately that schools fuelled societies, not just economies. De Botton hadn't. When he couldn't change the world through politics, he'd decided to mobilise a phalanx of East London kids to do so. We talked about Saul Alinsky, the community-organising guru who had inspired Barack Obama.* Year 11s at the school studied his techniques. The school was built around the idea of craft. It was a route to fulfilling our individual and shared humanity.

While de Botton was teaching, he'd read *The Craftsman* by Richard Sennett, a Chicago sociologist, thinker, smoker and typer. Sennett wrote that though it may 'suggest a way of life that waned with the advent of industrial society, craftsmanship names an enduring, basic human impulse, the desire to do a job well for its own sake'. Sennett referenced his teacher, Hannah Arendt, and hers, Martin Heidegger, warning that a technologically driven society threatened a 'lost space of freedom'. In an analysis of the NHS, Sennett perceived the finely honed craftsmanship of doctors and nurses – comfortable with

---

* Born in Chicago in 1909, Saul Alinsky was a community-organising genius. After working initially to organise the Back of the Yards neighbourhood in Chicago, which had been the topic of Upton Sinclair's *The Jungle* due to its horrific working conditions, Alinsky spent his life working to help America's poorest, most voiceless communities improve their living conditions, by organising them to stand up to powerful interests. He went on to start the Industrial Areas Foundation, whose community-organising approach is standard practice for labour unions, living-wage campaigns and grassroots political movements to this day. His *Rules for Radicals: A Pragmatic Primer for Realistic Radicals* (New York, Random House, 1971) is a bible of organising, beginning with these lines, 'What follows is for those who want to change the world from what it is to what they believe it should be. *The Prince* was written by Machiavelli for the Haves on how to hold power. *Rules for Radicals* is written for the Have-Nots on how to take it away.'

ambiguity, moving slowly, undertaken in a spirit of inquiry – being worn down and crushed under big data, KPIs and routinisation. Instead, he argued, 'the enlightened way to use a machine is to judge its powers, fashion its uses, in light of our own limits rather than the machine's potential'. This was the essence of craft, using tools to follow a human purpose.[22]

'I feel like we're selling voodoo,' said de Botton of the school's approach. I was buying it.

A mixed three-form entry, all-through school of 800 Newham kids, School 21 had opened its doors five years earlier. De Botton and Hyman felt that academic success had come to dominate schools at the expense of character, well-being, 'ideation' or problem solving. They intended to rebalance learning around the head, heart *and* hands. Kids would excel at English, maths and science – they'd be judged on these whether they liked it or not – but beyond that they'd do things differently. Project-based learning would be the norm, in which kids might learn history through drama or art. There would be a real-world element, with Year 10s doing six months of weekly half-day work placements, and other kids undertaking community projects. They'd aspire to make meaningful works, publishing books or creating sculptures. Instead of exam coaching, kids would engage in community-strengthening circle time.

In the drama studio, I'd had a first glimpse. A bright, high-ceilinged room, it was plastered with slogans, posters and newspaper front pages, like Mark Wallinger's Turner Prize-winning re-creation of Brian Haw's Parliament Square protest banners. On soft stools, ten girls and boys in purple V-necks and black blazers sat in a circle debating the impact of social media on talk. Ten observers stood shadowing them, making notes about their speeches on clipboards. Mr Shindler paced among them in a blue roll-neck and glasses, prompting, challenging and correcting like a particularly exacting theatre director. The class was Oracy. Year 7s were learning the craft of structured talk.

'Farooq, how could you toss that into the circle?' asked Mr Shindler of one of the boys. 'Provoke, challenge the circle.'

Each child had been assigned a different role in a group discussion. One played a parent, while another played a student. Others were teachers or social workers. They were discussing the effect of technology on kids' ability to have conversations.

Farooq, representing an app company, addressed himself to a pretend-parent. The lack of talk at home wasn't down to the devices, but the adults.

'If you interact with your kids more, there is more chance that they'll talk with you,' he offered. There was a pause.

'Look at your stems,' hinted Mr Shindler. 'Is there a way of provoking or challenging those that haven't talked?'

In a craft, you aimed for expertise. Step one required mastering a set of prescribed skills. In their hands the children clutched schemas that broke talk down into the physical (how we are able to use our voice), the emotional (how we make an impact on our audience), the cognitive (how we make our arguments) and the linguistic (how we express ourselves). Every student had sentence prompts to get started in conversation. They were now practising with them. Mr Shindler kept the intensity high, ensuring the practice was deliberate, coaching them on their performance. In a few years they'd master these basics and become fluent speechmakers. It was pure Bloom.

'I never dreamed I'd be making kids stand up in Year 7,' whispered Mr Shindler. In most schools they didn't. 'I thought it would kill 'em.' Instead he'd been wowed by their capacity to feel 'comfortable with the uncomfortable'.

Earlier, I watched as he'd coached them in their speeches. Learning at School 21 was arranged around six pillars: eloquence, grit, craftsmanship, expertise, spark and professionalism. Translated for the kids that meant speaking fluently and with confidence; always giving 100 per cent; creating beautiful work; practice makes perfect; always asking *what if?*; and being ready to learn. The pinnacle of the approach was 'Ignite', which the Year 7s were preparing for with Mr Shindler. Annually every student in the school gave a TED-style talk in front of a large audience of parents and peers. It demanded eloquence, craftsmanship and expertise. Year 7s prepared talks five minutes long on a topic of their choice. By Year 11, you were expected to go full TED.

'That's a list. *Craft* it,' urged Mr Shindler to Alistair, whose speech on President Trump was just a timeline of events.

Around the room, the children sat typing or writing out redrafts of the text of their speeches. Here was the craft.

The guidance Mr Shindler provided was like TED speaker prep. He'd asked them for a hook, anecdote, humour and seriousness in balance, a climax; pushed on quality of language (clever, original, playful, patterns and rhythms, metaphors, alliteration, different tones); suggested dialogue, characterisation, motif, painting of different shades, facts, quotes, props, visuals. I thought it might be overwhelming them.

'Ask Mum,' he said to Farooq, looking to improve the speech's relatability by rooting it in his family's story of migrating to London, 'why did you come here?'

You didn't get good at something only by thinking or reading about it. Expertise came through practice. The rest of the school's approach was built around this principle. In long 75-minute lessons, the kids worked on projects. History and drama class for the Year 9s were combined into a single unit to dramatise London scenes from the Second World War. Another class was bringing the Yalta Conference to life, playing Stalin, Roosevelt and Churchill, and dramatising the imminent clash of civilisations. I wasn't sure how it would help at GCSE (though the school's first Year 11s would achieve well above the national average in 2017), but then that was the point. In the Sixties, academic Marshall McLuhan had coined the phrase 'the medium is the message'.[23] It wasn't the content projected through a new media that was significant, but the *form* that media took. Less significant than the existence of PewDiePie or cute animal videos was the fact we were all now glued to our phones.

For schools, the medium was the message too. Many claimed to take creativity seriously, talked a good game about kids learning to collaborate. But usually you'd see the youngsters with their heads down in books, spending their time mastering core knowledge and skills. If you wanted kids to learn creativity and communication, they had to *practise* creating and have space and time to communicate. I was struck by the ease with which kids at School 21 took to a wide variety of activities without the usual adolescent embarrassment. They acted, made speeches, discussed, debated. They even danced in front of one another, something unimaginable from my stiff and formal schooldays. In all, they seemed well set for the world around them.

'Imagine if you wanted every child in the school to be a published author,' said de Botton. Freed from the constraints of exams he pictured a school in which kids graduated having done a TED talk, published a novel, performed in a play, completed a piece of scientific research, even run an election campaign. Not only were they practising real world skills, but their range of experiences meant kids had a much better chance of finding their *métier*, something denied to Lilas Merbouche and many others like her. I thought back to the Oxford Martin Report. Complex communication, creativity and large-frame pattern recognition were the human qualities we had to cultivate. School 21 was taking these seriously. But although creativity was a tool for productivity and finding a place in the economy, for de Botton it was even more importantly a route to purpose.

'I genuinely believe our children are looking for purpose. And schools are places where people come together to discover purpose.'

The world was changing too fast to be dogmatic. It was important to be open to new ideas, to welcome challenge to your worldview. School 21 was building this into its ethos – and its kids.

'Ten years ago, no one would have predicted the growth in hi-tech jobs, would they?' said de Botton. He was betting on his kids being prepared for the unknowable world to come. 'I know "the jobs of the future haven't been created yet", but presumably it's unarguable that there are new industries.' In his view, these Stratford kids would be the creators of that new world. They wouldn't be scrapping for the last remaining positions in the status quo, but finding new ways to build a world.

'Schools may fail to provide the tools to do good work,' wrote Sennett in *The Craftsman*. He thought craftsmanship was our highest human purpose. It was at the heart of all that the kids of School 21 were doing. As I headed back to the train, I looked around me again at the faces from all over the world, the tall Olympic buildings, the prosperity and the poverty. The kaleidoscopic mix seemed to encompass the whole world.

The point was to change it, de Botton had said. It was the work of generations, but it started like School 21 had, with someone hacking the system. That same impulse had inspired Wildflower Montessori back in Cambridge, Massachusetts, and before that the celebrated MIT Media Lab.

The White Spaces In Between

I stood with Philipp Schmidt in a vast bright atrium on the sixth floor of the MIT Media Lab. We were staring at a locked glass door. Beyond, a roof terrace overlooked the picturesque domes and quadrangles of Cambridge, Massachusetts, deep in January snow. The door, he explained, was emblematic of an idea core to the Lab's ethos, the Hack. 'Hacking culture at MIT started with people unlocking doors to rooms they suspected held something interesting,' he said conspiratorially. In this case, a student had entrusted Schmidt with the secret that by slipping a thin strip of plastic down the side of the frame you could trip a motion-sensor installed to prevent lock-outs and get outside onto the open-air terrace without a key. The temptation of spending those long summer nights out there with the stars above the city was too much. Someone had found a hack.

Schmidt was a digital age Don Draper, toting a dark gunmetal-grey MacBook Air and worrying that his standing desk was a little douche-y. His politics, however, were Wikipedia rather than WASP. He was the Lab's director of Learning Innovation, and community, transparency and learning were his passions. The year before he'd arranged a symposium on Forbidden Research, inviting a celebrated academic to share what she'd learned at MIT. 'Her most meaningful experience was putting the fire truck on top of the dome,' he said. It was a legendary MIT hack. Over the course of a single night, she had orchestrated a team of forty undergraduates to assemble a one-ton fire truck on the roof of the huge neoclassical dome that dominated the centre of the campus, a massive undertaking. 'She's a post-doc in Germany now and she said that the thing she's most proud of in her academic career – the thing that taught her the most and where she's learned the most – was running this hack.'

I'd come to make my offering to the man-gods of techno-humanism. The Lab was a temple to twenty-first-century creativity.

I believed.

The place sat self-consciously at the edge of the possible. It was overwhelmingly new, every detail designed with learning in mind. A glass cube arranged over six floors, with the atrium up in the eaves and the director's office tucked away on the second floor, the Lab tried to re-create the ethic of an earlier MIT space, where all of the most fruitful interactions had happened in the communal areas as people congregated to exchange, talk and share. It was a maze. Each floor spread around an empty central space that was difficult to navigate and inefficient to get around. Staircases missed out floors. The lifts were hidden, as were the toilets. It was based on an insight by Steve Jobs. Usually buildings did a great job of separating people, promoting isolation and user-friendliness. But frictionlessness was the enemy of creativity, as it was to learning. You needed bumps in the road, like Christof van Nimwegen's hiccups. The building was desirably difficult.

'Beyond VR and Self-Driving Cars', read a sign. We were in the future's future.

The Lab was home to 26 distinct research groups. Fiefdoms, Schmidt called then, 'kind of like Germany before it became Germany'. Each pursued investigations around a mysterious-sounding theme. Affective Computing. Camera Culture. Synthetic Neurobiology. Four of them were housed on each floor in two separate lab spaces. Part of the design cast unlikely pairings together, just to see what might come of it. In Sculpting Evolution we met a glassblower from the Royal College of Art in London whose desk was strewn with abstract forms, 3D-printed coral and a 'wearable' glass lung of hundreds of microscopic twisting tunnels through which microfluidics might flow (the Lab had the world's first 3D glass-printer). Schmidt shrugged. Sometimes the point was simply to imagine. The purpose could be discovered later. On the other side of the room the Opera of the Future team (it wasn't a metaphor, they were imagining what opera might be in the future) was developing scenery and costumes from chitin, a bounteous naturally occurring material found in fungi and the shells of crustaceans. A corner display case housed interconnected staircases angling off in all directions. Escher as architect.

'There are so many ways you could talk about this,' said Schmidt. Sometimes talking was the point too.

'Making sure that people here bump into each other more and talk to each other more is more important than in a traditional academic environment,' he added. The Lab believed creativity resulted from interaction. People had a multiplier effect.

On a screen I watched a short movie by Mediating Matter. It foresaw a future nano-material that behaved in the real world exactly as a virtual material behaved in the virtual world. A red sphere appeared around which a line was drawn with a specially designed stylus. A double-click on one of the hemispheres prompted that semi-circle to disappear. Another double-click and the hemisphere duplicated itself to form a pair of scooped-out bowls that looked like egg-cups. Another point was that once you'd *imagined* something, then it became no more than a problem to solve. Diverge then converge. The Lab's speciality.

Founded in 1985 by two scientists, Seymour Papert and Marvin Minsky, the Lab was a technological point zero, source of much of that which smoothes our lives. The Lab's boffins had invented touchscreens, now ubiquitous in our smartphones; the GPS we use to navigate; e-ink, which fuelled Kindles, but was also making possible molecular printing and production of synthetic genes; wearables that did things like monitor your heart rate; robot prosthetics for amputees. They knew something about invention, about operating at the edges of things. It required bringing people together in diverse groups (all the non-elephant animals, said the Lab's current director, Joi Ito) and cherishing chance encounters. It involved both imagination and hard work. Helpfully, they'd codified the Lab's lessons in creative learning into four simple insights: projects, passion, peers and play.

'We're the huge white space between the clearly defined disciplines,' said Schmidt.

Recently, Ito had convened a meeting to answer the question 'What is the Media Lab?' Born in Japan, Ito was a 50-year-old this-is-for-everyone internet utopian, successful tech entrepreneur and former club DJ in Chicago. He was also a serial college dropout whose recent book, *Whiplash*, vaunted 'learning over education' as one of nine 'ways to survive a faster future'. But trying to tie the Lab down had been a fool's errand, like trying to measure creativity. In the end, they'd managed to agree on basic criteria to evaluate the output of the various groups. Was it unique? Did it have a positive impact on the world? Was it *magic*?

In one of the group spaces, I found Kim Smith. An eminent visual artist from New York, she'd been part of Sep Kamvar's team that had launched Wildflower Montessori. Their inspiration came from studying cities. 'Everyone loves cities,' she said with a smile. After developing a set of tech tools to map urban life, the group noticed that 'cities are very organic', and asked how they might build on that understanding to improve how they worked. Counter-intuitively, they found that manmade interventions often undermined the natural productivity of the metropolis. Highways actually stopped people getting around efficiently. Enclosing public spaces trapped people indoors. Shipping kids off to massive factory schools alienated them from their communities. The group had focused on bicycles, parks and schools.

'A common thread in all of these projects is trying to increase social interaction and reinforce community,' Smith explained.

The philosophy seemed to be to start with what worked then figure out how to multiply it.

'It was very messy. But it becomes less messy. It makes itself clear.' Wildflower emerged from a set of *what if?* questions. What if we were to design the school of tomorrow? What if there was a school on every corner? They'd felt their way into it. The school would be small, within walking distance. It would feel like the heart of the community. Smith herself added the idea of computational thinking, hand-crafting the wooden dot-matrix screens that the young punk had been playing with when I visited. On her desk lay an array of beautifully designed wooden trays, each made for kids to explore some aspect of logic or programming. Now that the group was wrapping up, she was launching a company to bring the materials to schools all over the US.

'We'd do things, mess up, fail, figure out what works.' I still wasn't convinced that Wildflower would one day be for everyone, but I was no longer writing it off. 'When something is working you feel it. When something is not working you feel it.' It seemed right that everyone should have the chance to fail and the space to create. Smith herself had been to quite a stifling school. Her parents were both high-achieving doctors, but she'd started going to an art summer camp and it had taken off from there.

'It's something you find,' she said. 'Everyone's creative.' I thought they could be. 'What school does in a worse-case scenario is it stifles that or deadens it.'

Explore Then Exploit

I returned to find Schmidt on the top floor. On the wall by his standing desk were Post-it note self-portraits of everyone who'd visited him at the Lab. I scrawled my own for the collection.

'If the book doesn't work out,' he said as I gathered my things, 'you can always get a job making cartoons for the *New Yorker*.'

On the landing an outsize cartoon cat built from orange Lego marked the entrance to Lifelong Kindergarten. The centre of learning at the centre of learning. I found Mitch Resnick leaning intently over the laptop at his desk. A physicist by training and a former science journalist, he had been drawn to the Lab after hearing Seymour Papert lecture. A leading global expert in artificial intelligence, Papert was also a learning theorist who coined an idea of constructionism, whereby kids learned through developing schemas, like those at Pen Green, on top of which they added new knowledge through discovery. The inventor of Logo, a green triangular turtle used to teach kids programming in the early Nineties, he had envisioned immersive online worlds in which children could learn.

In a fictional 'Mathland' kids might acquire maths in the same way they learned English in England.

Resnick's philosophy marked him out as Papert's spiritual heir, as did his wild hair and dark beard. His shelves were strewn with relics of a life in computer science. Among the science trophies, oriental beckoning cats with their one ever-waving arm and Lego Mindstorms kits sat a small six-brick Lego duck. Schmidt had told me about an online challenge that called for makers to combine the six bricks in as many ways as possible that still resembled a duck. The record holder had run out of steam at 195.[24] The remainder of the wall was taken up by hundreds of books. Their titles – *Out of*

Control, We Feel Fine, Emergence, I am Strange Loop, The Art of Human Machine Interfaces, Predictably Irrational, Alone Together – suggested that he saw as well as anyone the future facing our kids. We faced a familiar paradox.

'Technology is both causing the situation where creative thinking is more important than ever before, but also, if we use it appropriately, and it's a big if, it prepares people with new opportunities to flourish in an era of rapid change.'

Resnick's break-out success had been Scratch, a free-to-all online coding tool with which kids could program interactive games stories and animations, remix and learn from one another's efforts. Twenty million kids in more than 200 countries were using Scratch each month, learning to 'think creatively, reason systematically and work collaboratively'. Every day 20,000 new projects were shared online. It was brilliantly simple. Kids stacked commands in an online window in the same way they'd stack Lego blocks in real life. I'd given it a go and hadn't been baffled. Creativity was the thing.

'It's always been desirable and it will be increasingly vital,' said Resnick. 'I feel that people will always be able to do creative things that machines can't do,' he added. 'As long as people develop their creative capacities there will be a place for them.'

Schmidt thought the question that was always hanging around was 'What does it mean to be uniquely human?' It had to do with self-expression. In any act of creation, you attempted to convey meaning to another. This was how Resnick saw programming. Learning to code meant 'developing your thinking, developing your voice and developing your identity'. Like Nicolas Sadirac, he thought it was akin to writing. As you put words on paper, your ideas evolved. 'The computer expands the way people can express themselves.' For Resnick, creative self-expression was the object of learning. He'd coined the four Ps of Projects, Peers, Passion and Play that defined the Lab's ethos. After Scratch, he'd founded Computer Clubhouses, where kids learned by those principles.

'Schools are not set up to help foster creativity in general,' added Resnick. His self-ascribed mission was to break down barriers, of space and time, and between the disciplines as they were doing in Finland. But he also sounded a note of caution. Rigour did have an important place in the learning equation. 'It's important not to think that creativity is opposed to systematicity.' He was sad to see how Anders Ericsson's ideas about deliberate practice had spread. It wasn't that he disagreed with the importance of repetitive practice, rather that before getting to that stage you had to have the space to find what you were passionate about. 'Only when you're following your passions are you going to be willing to put up with all the repetition.' The best creativity, he thought, came from 'combining imagination and systematicity together'. I agreed. He had his own phrase for it: explore and exploit.

'The problem is, in the service of the rigour a lot of times the imagination and creativity gets squeezed out of the system. But it would be equally wrong to just embrace chaos, without any type of rigour.' We required a better balance.

George Bernard Shaw once remarked that where 'the reasonable man adapts himself to the world, the unreasonable one persists in trying to adapt the world to himself'. All progress therefore depended on the unreasonable man. I wondered how schools, bound by rules and traditions, could nurture this unreasonableness. The Lab was succeeding. You had to walk the borderlands between structure and freedom. That's where creative learning lived. It was the question the Finnish teachers had obsessed over.

'It has to still feel risky and dangerous, or you don't get the benefit of hacking,' Schmidt had said up on the roof.

It could only feel risky if it *was* risky. You couldn't fake it. At MIT things did go wrong. In 2011, the internet activist and computer-science genius Aaron Swartz had been arrested by MIT police having broken into a server room on the main campus to hook up his laptop and begin the systematic download of the entire JSTOR archive, the private digital library that stored the world's learning in the form of academic journals, books and primary sources. He planned to make available to everyone the back issues of all academic publications, believing nobly that knowledge should be for everyone. Federal prosecutors disagreed, charging him with wire fraud. Faced with a potential maximum sentence of 35 years' imprisonment and up to $1,000,000 in fines, Swartz committed suicide at his Brooklyn apartment in 2013.

The price of adapting the world to yourself was that it could resist the change. Attempting progress risked radical failure.

The elimination of this risk was the greatest threat to human creativity. A French philosopher, Alain Badiou, argued that everything meaningful, everything innately *human*, depended fundamentally on the possibility of failure. He lamented a world in which risk was removed by authorities that felt they knew better, by school administrators, even these days by algorithms that knew us too well. What space was there then left for failure, for creativity, for making a decision at random in a chaotic universe and then building meaning from it?[25] When it came to learning, we were so focused on ensuring kids mastered basics, which they really did need to, that we'd squeezed out risk-taking from their experience.

We now knew we had to put the hours in, Resnick's 'exploit', but we weren't allowing enough time to explore. I pictured Mervi's movie-makers out on the climbing frame with their video cameras. What they'd been doing hadn't seemed like *learning* to me. There was no lesson objective, no real rigour. Now I understood that this was the point. The medium was the message. You learned to write by practising writing. You learned imagination through practising imagination. Craft meant grafting to master core knowledge and to perfect the foundational moves and techniques of your field. But it *also* meant cultivating a spirit of exploration, having the freedom to fail. Grade nine piano geniuses rarely went on to become the greatest composers. Spelling bee champions didn't write great novels.

MIT Media Lab was full of first-class honours graduates, but it was run by a serial college dropout.

'It's really disheartening how few places like the Media Lab there are in the world,' said Schmidt. Any learning that took place at most universities was largely by accident, he thought. It was true of much learning in schools too. At the Lab, learning was by design. It embraced the principles of the lifelong kindergarten. 'It's accepted that kids play in the playground,' Schmidt had said, 'but when adults do it, it's not serious.'

I agreed about the adults, but not the kids. We no longer *did* accept that they played. Not when they were at school.

'We want everyone to be able to follow their own dreams,' said Resnick. 'But you also want them to be able to *realise* their dreams.' This was the paradox. 'If you just focus on rigour, you don't have the dreams. But if you go to the other extreme, you won't be able to realise your dreams.' I agreed. But I also thought that the rarefied environment of the Lab had divorced its inhabitants a little from the world at large. *Out there* were kids with the odds stacked against them, environments set up to make failure close to inevitable, young people who'd given up on learning. Wasn't the real world close to the opposite of MIT's idealised learning environment?

In the open-plan office space I paused by a cauldron of Lego bricks. It was clear that doing better meant ensuring that each of us found our purpose, discovered our *métier*, our craft. We *could* grow creativity in our kids, but only if we gave them space to play and to fail, then guided them through the arduous process of becoming expert. 42, Scratch and the Media Lab suggested how kids might learn to use the tools of tomorrow. But I worried that the vision was too tech-centric. In the future, our most valuable endeavour would be that of developing our human potential. Cultivating our ability to learn, to find meaning, to create and to co-operate fell to our teachers. Whereas for too long we'd thought of them as knowledge experts, I thought instead that we should see them as masters of the ultimate craft. In places as diverse as Seoul, Helsinki and New Jersey, I'd already had a glimpse into that reality, as you'll see in the next chapter. If we were born to learn, we were also born to teach.

Losing myself among the raised walkways in an attempt to find the exit, a phrase of Resnick's played over and over in my head.

'It works at the best kindergartens,' he'd said of the creative learning approach, 'it works at the Media Lab. We just have to change everything else.'

CHAPTER 6

Class Act

Masters of the Learniverse

If you can read this, thank a teacher.

Bumper sticker

The Robot Teachers aren't Coming!

Kim Su-ae looked bored. It was the first day back for the Korean language teacher at Dongpyeong Middle School in Busan, a large port city at the tip of South Korea. She paced slowly at the front of the classroom reading aloud from a textbook, from time to time correcting a student – 'Lift your head, Jun Seong' – or smoothing her red summer dress. A heatwave had pushed temperatures to 35 degrees centigrade, and the teenagers, hunched over their desks in white shirts and grey trousers, fidgeted to stay awake. A few gainfully followed along. Soon one, two, *three* of her 20-odd students laid their heads down and slept. 'She only sticks to the book,' said a boy. 'We get bored and fall asleep.' Su-ae's temper frayed and she sent Jung-min to stand at the back.

'I'd love it if every student in the classroom was focused and had the urge to learn,' she said afterwards.[1]

I knew the feeling. At Walworth, I'd struggled daily to motivate my students, to get them to concentrate for just a few minutes. While other teachers were conductors, choreographers, learning designers or data analysts, I was playing Whac-A-Mole. It certainly wasn't as simple as declaiming lines by Chinua Achebe before discussing.

'Find your identity as a teacher,' my tutor had said – unhelpfully, I thought – after a harrowing Year 8 class in which Kemal, David and Aleesha had been sent out twice each for refusing to work. But the mysterious *juju* of my inner teacher eluded me. Instead, I clung to scraps of structure, using the GCSE mark scheme to design learning and the Ofsted inspection criteria to model my teaching. Gradually, kids made a little more progress and my class became a little more soulless. I was frustrated. Colleagues encouraged me to keep trying, it would 'come with experience'.

I observed Ms Toworfe's fantastic Year 8 class, but *being* Ms Toworfe was hardly realistic. There had to be a method that worked.

In Busan, Kim Su-ae stuck with the textbook. It was a strategy, but it was sucking life from the class and depressing her. Even the feeble capabilities of the NAO robot teacher I'd met that week in Seoul extended to rote reading aloud of curriculum materials and preprogrammed questioning. The grading of Korea's standardised tests was already digitised. What purpose did Su-ae serve? As a teacher she was highly respected, but she was miserably confined in her work and couldn't engage her students. 'I don't know how to make that happen,' she continued, tears welling. 'Maybe I'm the problem.'

I didn't think that she was. Nor had I been. Instead, we had a problem with how we thought about teaching. In keeping with our outmoded ideas of learning, we'd long seen it as a *knowledge* profession, where subject expertise mattered most. I thought that doing better meant reimagining it as a *practice*. If we did so, we'd no longer assume that you'd best learn to teach simply by working it out for yourself, as I had failed to do. Nor would the profession consist merely of supervising kids as they followed along in the textbook, reading out predetermined lectures like Kim Su-ae. That was to automate teaching, to fit a highly capable human expert in as an unthinking cog in a leviathan learning machine.

If learning was the defining characteristic of our *species*, teaching was our most vital craft. This chapter takes that premise as its starting point. In the pages ahead we'll see two distinct answers in two very different countries to a simple question. What does it mean to teach well?

What Doctors Can Teach Teachers

We begin with a conundrum. What *is* great teaching? Picture your stand-outs. I see Mrs Taylor, the Scottish battleaxe at whose desk I used to read aloud in Year 2 or Mrs Midgeley, who taught me classics – yup, *latinam didici* – at Emscote Lawn aged 11. Later, I think of Mr Barlass's energy and wide-mouth-frog joke, Mr Morgan's indoctrinatory Marxism – death to beagles! – and the long discussions of *Hamlet* and *The Waste Land*, Jane Austen and George Eliot with Mr Hall. I liked going to their classes. They had charisma, knew how to captivate. I sensed I was learning important things. But they weren't carbon copies. Each had a subtle individual genius that resided in who they were. They'd accessed that mysterious magic that meant they could engage and inspire other humans. It defied analysis. There could be no science to judging teacher quality. Every teacher was great in their own way. It's why teachers were *born*, not made.

That's exactly what they once said about doctors.

In 1846 a young Hungarian doctor called Ignaz Semmelweis took up the position of Chief Resident at the First Obstetrical Clinic of Vienna General.[2] The clinic was one of two at the hospital that provided infant care *gratis* and should have been a haven to underprivileged women. Yet its extreme maternal mortality rates – 10–20 per cent of mothers died of puerperal or 'childbed' fever – meant women begged instead to be admitted to the Second Clinic, whose mortality rates were three times lower. Semmelweis was deeply moved and began a meticulous inquiry into practices at the two clinics, gradually eliminating possible factors. Medical techniques were identical at both, as were the climate and religious practices, with the Second Clinic actually *more* overcrowded than the first. There was just one difference: the First Clinic trained medical students, the Second Clinic midwives.

Semmelweis's eureka moment followed the death of Jakob Kolletschka, a colleague at the First Clinic, who was nicked by a student scalpel during a routine autopsy. Examining his friend's cadaver, he noticed a similar pathology to that which had been observed in the women dying of puerperal fever. He concluded that the doctors and medical students had been transporting 'cadaverous particles' from corpses into the clinic. In May 1847 he instituted a groundbreaking handwashing regime, with chlorinated lime used to destroy any poisons. The previous month had seen maternal mortality reach 18.3 per cent, but, as soon as students regularly washed their hands in the new solution, the rate fell to 2.2 per cent in June, 1.2 per cent in July and 1.9 per cent in August, superior even to the Second Clinic. It was a triumph of inquiry.[3]

Yet Semmelweis's story didn't end with that happy discovery. At the time, Austro-Hungarian physicians saw themselves as we see teachers today, as autodidacts and purveyors of a mysterious craft. Diseases like puerperal fever were deemed to result from things like dyscrasia or 'bad mixture', an imbalance in an individual's four humours. A doctor's 'expertise' lay in diagnosing which of blood, phlegm, yellow bile and black bile was out of kilter in a patient. Great skill was required in what was a highly complex task, given that each person contained their own individual cocktail, and after much deliberation an appropriate remedy would be suggested, such as bloodletting. These doctors ridiculed Semmelweis, refusing to take his practice of handwashing seriously. Medicine was too complex, too *human*, an art form to be improved by mere data analysis, or something as degrading as specific, concrete practices.

It is like learning today. In Semmelweis's time there were many brilliant, caring doctors. Just as there are tens of millions of inspirational, committed teachers worldwide. It doesn't mean we should avoid trying to get better. We still have our own four humours, solemnly labelling kids 'musical-rhythmic', 'visual-spatial' or 'bodily-kinaesthetic' even though the theory of eight 'multiple intelligences' has been disproved.[4] No, it means teachers, like doctors, have continually to ask how they can become masters of their craft.

In Newark, New Jersey, a radical new programme recently decided this meant training novice teachers a little more like new doctors. A little more, in fact, like athletes.

Little Things Have Big Muscles

Da'jia Cornick performed *alert* at the front of her sixth-grade maths class, back straight and eyes flitting from face to face. Five feet nothing in blue suede shoes and barely older than they were, she was *all* business. The students, clad in green polo shirts and khaki trousers, shuffled up in their seats to mirror her poise. A few hands were up already, but she wanted more. They called this 'Wait Time'.

'Track Ignacia,' she said. Twenty-five heads turned to face a dark-haired girl in the centre of the class. The class was arranged in three neat rows of pairs, workbooks open on desks. All eyes were on the speaker.

'You multiply by thirty,' she said confidently. Ricardo had given an incorrect response to *How do I get from four to 120?* The other students were now assisting him. This was 'Break It Down', the process of exploring wrong answers. 'No Opt Out' meant that Ricardo was still expected to give the right answer – after a little help from his friends. He listened to Ignacia then repeated her answer in his own words.

'You got it, kiddo.'

Cornick dipped her clipboard under a mounted camera, projecting the next problem on the whiteboard.

'What would I get if I simplified that?' She planted herself in front of the class and indicated a problem: 25 over 100 equals *x* over 120. Holding her pen like a conductor's baton, she repeated herself slowly. 'What would I get if I *simplified* that?'

Ricardo was thinking hard. A cluster of hands went up.

'Oh, I love that back row. *Killing* it.'

That was 'Positive Framing', specifically the sub-skill of 'Narrate Positive' designed to boost the 'J Factor'. 'J' stood for 'joy'. She now peppered them with problems, quiz show fast.

'*Track* Ashley,' she continued. Heads turned and Ashley answered. Cornick ping-ponged extensions around the room – *Track Tony, Track Manuela, Track Tiffany* – without pausing for breath. The kids were sharp, ready for it. If an answer wasn't 100 per cent, Cornick pushed back until they nailed it. 'Right is Right' meant accepting only perfect responses. Kind of right or mostly right would be to indulge in the soft bigotry of low expectations. She was tuning up her youth maths orchestra to be virtuoso performers. Satisfied they knew their parts, she set them to rehearse a simplification symphony.

'Scholars,' she called, lifting her baton-pen. 'Pencils *up*!'

Twenty-five pencils joined hers, four a little half-heartedly.

'I'm sorry, *scholars*,' she chided, noticing some crooked arms. They would 'Do It Again', repeating it until they all got it right. It was another example of her high expectations.

'Pencils *up*!' She scanned the room. 'Strong vertical,' she reminded them, exacting extra professionalism. 'Stronger.' Arms and backs reached high.

'Alright! Continue working.' She hit start on her stopwatch, and they got down to simplifying.

Beside me Jamey Verrilli was coach-pleased. 'When she asked a question, she pushed for 100 per cent of hands up,' he whispered. This was *great*. It was especially exciting for Verrilli, who'd been Cornick's head teacher when she was a young student at the same school we were visiting, North Star Academy, named after the abolitionist newspaper founded by the nineteenth-century civil-rights activist Frederick Douglass. Verrilli had co-founded North Star in 1997 with serial education innovator Norman Atkins in an effort to begin to close the achievement gap in Newark, New Jersey. America's second-poorest city, the desolation of its surrounding areas had long been sung by its most famous son, Bruce Springsteen. Two decades later Verrilli was overseeing the first year of Cornick's teacher training at the Relay Graduate School of Education, where she was a Resident and he Senior Dean. He was a maestro and an activist, in a bow tie and Black Lives Matter pin, his fastidiously packed rucksack and pilgrim beard adding to the image of a social-justice frontiersman.

He was thrilled too about the exchange with Ricardo. 'You try it, you get it wrong, you get some feedback, you get it right.' You had to normalise error if all kids were to learn.

Cornick toured the classroom, clipboard prepped with an exemplar answer and the names of students most likely to struggle, timer ticking down. She briskly visited them in turn, while Verrilli interpreted. This was 'Aggressively Monitoring'. Her surgical attention to detail was all the more thrilling for the fact that it was only her fifth month in the classroom.

'Give Rodrigo a dollar!' she whooped over the gentle scratching of pencil on paper, 'He's rocking it.'

Later Verrilli explained the Relay approach. At its heart was the idea that teaching was a craft. This was why I had come to visit. Whereas other training programmes invested heavily in theory – talking and writing about teaching – Relay cared above all about *practice*. The moves Cornick used so seamlessly came from *Teach Like a Champion*, a revolutionary book by a teacher, principal and Harvard MBA, Doug Lemov.[5] He'd been a struggling novice educator in a poor part of Boston when a colleague had offered him some simple advice. If you want kids to follow directions, stand still while you give them. It turned out to be effective and it got Lemov thinking. Perhaps teaching wasn't voodoo dependent on some innate talent, however much it felt like it with Year 8s. Maybe there were other skills like this. As a keen soccer player, he knew that clear and specific instructions like 'close the space' or 'narrow the angle' led to improved performance. If excellence on the pitch meant mastering an array of intricate, specific skills, might it not be the same in the classroom?[6]

When Lemov later became a consultant to failing schools, he brought this question with him. Frequently he found teams of teachers working tirelessly to improve, but short on ways to get better. There still seemed to be a taboo around defining great teaching. He started to take his search more seriously. In 2002 the No Child Left Behind Act had furnished US administrators with reams of data on school performance. He began to sift through it, looking for outlier institutions where low-income students achieved against the odds. North Star Academy – where Verrilli then went by 'Principal V' – was first on his hit list. The students were exclusively African-American and Hispanic, nine in ten received free lunch and just one in ten had a college graduate parent. Yet a typical North Star eighth-grader outperformed the New Jersey average on state tests by more than 30 per cent in English Language Arts and more than 35 per cent in Math. The school didn't just close the achievement gap, but *reversed it*. Lemov persuaded a wedding videographer friend to accompany him on the visit.

Those first films of high-performing teachers sparked a five-year documentary project supported by Uncommon Schools, which Norman Atkins had founded with Verrilli's help. Hour upon hour of video footage was gathered, which Lemov then pored over with the keen eye of a soccer coach. He noticed that one thing united all great teachers – a knack of manipulating kids' attention, the holy grail of learning. They also held high expectations, wasted no time at all, and made learning highly engaging. But these truisms weren't going to help failing schools. After all, if someone told you to 'hold high expectations' what would you do? So he looked more closely, focusing in on specific techniques – moves – that the outlier teachers employed, giving names to what he perceived, like 'Sweat the Details', 'No Warning', or 'Cold Call'.[7] Great teaching would no longer consist of blood, sweat, tears and magic, but of blood, sweat, tears and a concrete and specific set of 49 micro-practices. 'Seemingly small, deliberate changes in teacher words and actions can, over time, produce dramatic improvements in student achievement,' wrote Lemov when his taxonomy was published.[8] Relay Graduate School used many of the methods as part of its training programme.

'I got one, two, three, four, five, six,' counted Cornick, looking out for hands. One child had her fingers crossed, a time-saving 'Seat Signal' that said she needed the bathroom. 'Let's pick it up! Seven, eight. How many questions did Riley actually complete?'

'Forty-eight,' said Lincoln.

'But the prompt asked us how many more did Riley complete than Annabel?' Verrilli winced. 'She's feeding them a little too much, but she'll learn,' he whispered. The improvement Verrilli thought Cornick should work on – 'it's kind of a 2.0 level' – was 'Ratio Part One'. Teaching aimed to maximise the cognitive load on the kids, but she'd just missed a chance to get them to recall the prompt themselves. For Verrilli, these details were significant – add them up and they were the difference between kids learning and not. Attention to detail – 'Sweat the Small Stuff' – was in the DNA of Uncommon Schools, as it was in Relay and had been at KSA in London. All lesson planning was done centrally to maximise quality and save teacher time. Cornick was teaching *exactly* the same sixth-grade maths class that was right now being taught in the 40 other Uncommon Schools in Rochester, Boston and Brooklyn. Likewise, the trainee teachers were all learning exactly the same set of moves. Practising your craft was everything, and it required a formula.

Cornick clapped twice. 'Clap-clap, clap,' came the response from the kids. Their pencils were down and every one of them was attending to her, as though under a spell. But it wasn't voodoo. Practice worked. And yet something niggled. Perhaps I was overly wed to the idea of teaching as an art, but Relay's teaching formula seemed somehow a little restrictive, a bit *too* formulaic. The undertones of relentless Tiger Parenting that produced the stilted über-linguists of the spelling bee world were strong. Wasn't there something a little, well, mechanistic about it all? Was this really how you developed craft?

Verrilli was giving a 'Deliberate Practice' class that afternoon to a group of Residents. I hoped there that I'd find out more.

The 2 Sigma Problem

Teachers waited until 1985 for their Ignaz Semmelweis moment. It came again from Benjamin Bloom. In his studies of talent, he noted that the development of expertise also depended on having access to a great teacher, ideally a succession of them. A future virtuoso would often begin working with a local expert to cultivate basic skills, but would then typically transition to a nationally recognised coach to ensure their talent really flourished. Bloom highlighted an obvious problem with this as a way of thinking about the learning of all kids. National-level experts were thin on the ground. In the most successful training scenarios, they provided intensive one-to-one support to the prodigies. Noting that within the vastness of an education system it was impossible to replicate these conditions, Bloom set out to study how teachers might most effectively re-create the one-to-one effect of tutoring in classrooms of 30 students. He called it the '2 Sigma Problem'.[9]

Bloom devised an experiment to evaluate three methods of teaching: conventional (30 kids learned in the usual way), mastery (same number of kids, but with extra formative assessment to give them individual insight into how to improve) and tutoring (a dedicated tutor worked with one student, or at most three students). The results were dramatic. The kids being tutored achieved on average two standard deviations above kids in the conventional classroom setting, hence 2 *sigma*. In layman's terms, this meant that the *average* – not the best – kid receiving tutoring would outperform 98 per cent of those in a conventional classroom. It was possible for each kid to learn more, *a lot* more, than they were currently doing in the classroom of the average teacher. The findings explain how super-tutors in the UK or US today command as much as £1,500 for hour-long one-to-ones.

But Bloom had no interest in suggesting every parent go looking for a tutor for their kid. Instead he asked if it might be possible that kids learned *as though* they were in a one-to-one session. That was the essence of the teaching craft, a question inviting near-bottomless exploration, the starting point for an unending quest to do better.

In Finland, supposedly home to some of the world's greatest teachers, I'd met one who was searching at its furthest depths.

The Art of Teaching without Teaching

In the early dark of the north Helsinki suburb of Martinlaakso, the low-rise Brutalist units lining the path from the station were alive with graffiti. It was an unremarkable setting for a remarkable visit, to meet Finland's Most Famous Teacher, Pekka Peura. In his classroom, winter coats were draped on chairs, as were a couple of the 28 students, one or two of whom looked fresh from bed. Sat around square tables in groups of three or four, they gossiped and checked their phones. Others had already booted up laptops, plugging them into dangling ceiling sockets, and were hard at work. The atmosphere was relaxed and conscientious, the uniform of jeans, sweaters, trainers in black, grey or navy impeccably Scandinavian. Through the ground-floor windows, silver birch and pine stood ranked against the flat lightening sky.

At 8.20 a.m. Peura strode to the back of the classroom, shared a joke with students and kicked the door shut. Even more casual than the kids, he wore a band T-shirt, knee-length denim shorts and an unzipped school-issue hoodie. 'Scottish indy guitarist,' I scrawled in my notes. His blonde beard and helmet of hair sealed it. Peura walked back to the whiteboard, passing the cartoon of Albert Einstein – *always* Einstein – with a yo-yo, flipped on his computer and projected a problem on the screen.

It was a multiple-choice question. *After how many seconds would the car overtake the tractor?* (a) 3.0s, (b) 4.5s, (c) 6.0s, (d) 7.5s, (e) 8.2s.

Around the room, the blonde 16-year-olds gazed at the graph projected on the whiteboard. A few tapped on their laptops. Others peered into their phones. A widget on Peura's display tallied the responses. Twenty-one had blipped in an answer. I knew this approach and waited for Peura to reveal the answer. How many had answered correctly? What he did next came as a surprise.

'Turn and talk at your tables,' he instructed the students. How had they answered and why? The teens turned into their table groups and began talking purposefully in Finnish. After a few minutes of discussion, Pekka repeated the same problem on the whiteboard. *After how many seconds would the car overtake the tractor?* (a) 3.0s, (b) 4.5s, (c) 6.0s, (d) 7.5s, (e) 8.2s. Again, students considered the graph and beamed in answers from their devices.

Still Peura didn't reveal the answer. Instead, he now displayed the results of the first poll: 36 per cent of students had chosen answer (a), 27 per cent (b), 18 per cent (c), 5 per cent (d) and 14 per cent (e). Interesting, said Peura. It showed there were some people in the class who got it, and some who did not. Even more importantly, it suggested that among the kids there was a variety of misunderstandings – they were wrong for different reasons. After a bit of back-and-forth with the students, he then revealed the results of the second poll. 39 per cent had gone for (a) this time, but 50 per cent had gone for (b). No one had gone for (c) or (e) this time. And just 11 per cent for (d). They'd changed their minds. What did they make of it?

I was intrigued. I'd not seen it done like this before.

'I know you are going to answer wrong,' said Peura in the staffroom later on, between sips of bitter black coffee – the Finns' national drink. The fridge door was adorned with magnets bearing faces of hundreds of teachers, edges airbrushed like cheap religious icons of Baptist ministers in a Southern church. There were a couple of a clean-shaven Peura with a bowl haircut, from his *Hard Day's Night* era. The furniture was functional pine. 'That's the whole idea of this task,' he went on in flawless, clipped English. 'You answer wrong and then your brain might start thinking about, OK, there's something that I don't know.' Then you might start practising. When it came to physics and maths, kids' misconceptions often rested on an idiosyncratic internal logic. If the teacher simply told them the right answer, they would find no place for it in their understanding and the underlying cognitive architecture would remain unchanged. They had to think it for themselves. This for Peura was the whole point of teaching.

'If the teacher is in control of what they learn and when, they don't know how to learn,' he continued. Peura had built a huge following across Finland for his thoughtful approach to the craft of teaching – no mean feat in a country that took the profession so seriously. He'd turned the idea of teaching on its head. Rather than seeking the most effective approach to getting high school kids to learn maths and physics content, he'd asked instead, what are the conditions in which kids will learn the most? It was a subtle distinction. In the classroom, I'd wondered if it wasn't a bit slow. After the rigorous pace of KSA and Relay, it had felt a bit under-whelming. Were the kids getting through all they needed to? 'Well, of course, I *speak out* less content,' said Peura, 'but the key is how much the students are learning. Even if I try to teach 100 per cent of the content, it won't be 100 per cent that every kid learns.' For Peura, the differences between kids meant you had to find a way to differentiate the learning experience at all times. The deeper point was to understand the class as a group made up of individuals, then try to figure out a way for them all to learn as much as possible.

Back in class, Peura briefly revealed that the answer was (b) and had a couple of students explain why to the whole class. Then he projected a table. The left-hand column listed the dozen or so content areas that students aimed to master in this element of the physics course. The other four columns were headed with emojis. Kids were to self-evaluate their level of comfort with the topic, se-lecting either Strong Arm for 'I know this well enough to teach my friends', Thumb Up for 'I've got it', Half-Smile for 'I know some of this but still have work to do', or Face Crying With Laughter That You Think I've Got Any Clue About How To Do This. The students gazed at Peura as he gave the instructions, before turning back to their computers and getting on with the self-assessment. There was no long explanation of how to figure out the right answer, no effort to probe the mistakes kids had made. Peura left the students to do that among themselves. Ten minutes into class, and it seemed like his work was done.

'I don't write anything on the board,' he confessed. 'I know that they can find the information that they need if I just guide them to look in the right place.'

The classroom didn't exactly feel futuristic – some of the ideas were a hundred years old – but Peura's spirit of inquiry and a pursuit of the craft of teaching were radical. His aims too were different. He'd arrived at what global research has shown to be the highest-impact strategy for learning – student self-evaluation – and built an environment and set of tools designed uniquely to nurture it.[10]

I sat down next to Patrik as the students got to work. He showed me the windows open on his laptop screen. A graphic organiser held the same list of topics and emojis that Peura had projected. This was where Patrik kept track of his progress. For each of the topics, there were links to learning materials – which he'd access at his own pace – and tests, which he could complete whenever he wanted. 'You do it, you mark it, you move to the next one,' he explained. 'Then you evaluate yourself here.' He liked it: 'You can do different stuff at your own pace.' Occasionally he'd ask Peura for help, but more often he'd discuss with his classmates at the table. He had access to their self-assessments, and could look to see who in his class had selected Strong Arm, and might be able to teach him a topic he hadn't mastered. There was also a Googledoc open, where he was collaborating with the others at his table to create course notes. Each group of four did this, and then each set of notes was shared with the whole class. Peura wanted to build strong learning habits. His latest innovation was around teams.

'If feeling safe is the main thing in teams, then as a teacher that should be the first thing to concern ourselves with.' He'd been

reading about a multi-year study undertaken by Google to understand the criteria of the 'perfect team'.[11] The key, it had discovered,
was 'psychological safety'. In his class, kids had chosen their own
teams then written rules for themselves. He'd discussed it with them,
but the decision had been theirs, and the self-made rules were the
only ones they were accountable to – 'we are trying to learn' being
Peura's only classroom rule. He'd taken a similar approach to his
practice over the nine and a half years he'd been a teacher. His tinkering had begun seven years earlier when instead of giving students
a new homework assignment each lesson, he'd handed out a booklet
containing all of the tasks. 'It took two weeks. The first one came
to me and said, "Everything done."' He'd realised then that the
traditional approach wasn't really working for *any* kids. Too fast
for those that were behind, too slow for those that were ahead. He
started to experiment – how could he ensure that each individual kid
maximised their learning?

It was Bloom's 2 Sigma problem. His answer was to leave them
to the content and to concern himself with developing and providing
the tools kids needed to jointly manage their own learning experiences, and creating a motivational culture and environment that
encouraged their natural born learning instincts.

'That's an idea we should just *delete* from learning,' Peura told
me when I asked if he worried that some kids in his class might
be left behind, 'I think we should just delete the whole idea of left
behind.' It offended his mathematical mind – 50 per cent of kids
would always be below average – but it was even more troubling to
him as a human. He worried about the effect this way of thinking
had on students' self-esteem. 'It's normal to be different,' he said. But
we weren't taking that into consideration in our teaching. Instead,
we were treating everyone as though they were clones. A lot of this
came down to an insight popularised by an Australian academic
called John Hattie that learning, what's going on in your brain, is
invisible. 'It's easier to understand,' he explained, 'in playing piano.
And it is normal in sports.' There we could *see* someone's current
ability level. 'But in school in math, history, psychology or physics,
we're not allowing them to be different.' It was easier as a high school
teacher to see things this way, but the point was fair. You couldn't be

great – couldn't be really *expert* – at everything. By demanding they excel at everything, we weren't allowing kids to find those few things that they wanted to master. 'We should find the things the student is interested in, then invest in that.'

In Finland, all classrooms were mixed ability by law, one of the reasons the gap between the highest and lowest performers was relatively small on the PISA international comparison tests. He estimated that it would take about five years' learning for the students who were currently able to do the least to catch up to the current knowledge and skill level of those who could do the most. Peura knew most of his students wouldn't go on to study physics at university, though a few would – I'd spotted them on a table near the front – so he sought to grow their capacity as learners.

'I try to teach the people first. The physics is just – it's a tool – it's the content.'

People were valued highly in Finland. The teachers too. At lunch I'd been struck by how seriously they took their craft – but also how normal they all were. I was reminded that only one in ten applicants made it onto teacher-training programmes. Although Peura told me that it was easier for high school maths teachers. There were only seven applicants for the 25 available places at Helsinki University. Further, Finnish schools had none of the fragile hierarchy of those in the UK. There was a principal, but it was a basic administrative role. There were no layers of middle managers at all. Teachers took collective responsibility – and they seemed to like their work – so it made sense that the kids did too.

Peura pushed this further than any teacher I'd ever seen. In Finland, kids took no standardised test until they came to graduate from high school. In Peura's class, they awarded themselves their own grades.

'I just ask them,' he said. 'It's close enough.' Peura, like Saku Tuominen, thought the comfort we found in numbers as a way of evaluating learning was illusory, even harmful, coming at the expense of developing our own judgement. He preferred to grow his students' ability to self-evaluate. Were they accurate? I asked. 'Hard question,' he replied, 'what does it mean to be accurate?'

He showed me the tables that he'd created for students. There were three self-reflection templates for them to fill out. 'How well did you learn learning skills? How well did you learn social skills and self-studying skills? How well did you learn physics?' Each was broken down into a dozen sub-questions, probing things like *sinnikkys* – Finnish for perseverance and hard work – but also things like teamwork and content knowledge. They simply rated themselves blue, green, yellow or black for each. The ratings were the basis for an end-of-course discussion with Peura.

'How can they start learning themselves?' he said. This was the purpose. He'd deleted his own authority from the classroom, made the kids tutor themselves – and each other. Several studies had shown kids learned better from their peers than they did from a teacher.[12] 'The greatest effects on student learning occur,' wrote the Australian researcher John Hattie, 'when teachers become learners of their own teaching, and when students become their own teachers.'

It also built growth mindset. Patrik told me that before he'd joined this class he'd thought he wasn't great at maths. Now he understood – everyone in the class did – that there wasn't such a thing as being good and bad. It was where you are now, and where you're heading next. If, as Carol Dweck wrote, the future belonged not to the know-it-alls, but the learn-it-alls, then this was them.

'Our students have been nine years in school and they have *no learning skills*,' Peura had said. 'I try to teach them learning skills. At that point I don't have to teach them physics almost at all.'

If Relay developed maestro teachers who could captivate students, Peura was trying to develop maestro students who could captivate themselves. It was a subtle, capricious craft. It was an alternate route to expertise, beginning not with routine skills, but a passion for a field and a slow process of tinkering.

This was how Peura had approached his work. He was a master craftsman, a Zen master of learning, exploring the Art of Teaching without Teaching. The trick as he saw it was to unlock and nurture kids' love of learning, freeing them to invest their own energy and take responsibility at school. 'People don't hate learning when they are born. There have been some events that have made that so.' He'd recently read a report that at age 15 at the end of ninth grade, 40 per cent of boys and 35 per cent of girls in Finland didn't like school. 'I think it would be quite easy to make them like to be in school.'

He tinkered with motivational strategies and the use of basic technologies, like Googledocs. He wanted to set an example that other teachers could follow.

Heading out on the train that morning with the commuters in their brightly coloured puffas, I'd been reminded of a talk given by the artist Grayson Perry's alter ego, Claire. Creativity didn't begin with originality, she had said. Instead, artists picked a path based on something that they liked or were interested in and stuck at it for a while. It was like catching a train at Helsinki station. You had 12 choices, but no real way of knowing whether you were heading to Riihimäki, Leppävaara or Kirkkonummi. What was important was to get on – and stay on – the train. If you lost your nerve and returned to the start, you'd be stuck with that same difficult choice, only now you'd be late. If you wanted to succeed creatively, you should stay on your train. The early work would always be derivative, but eventually you'd find your route. It was all down to tinkering.[13]

'I have a rehearsal,' said Pekka, when I asked him at the door about his plans for the evening. I'd been right. Finland's Most Famous Teacher also played bass in a band.

Back on the train at Martinlaakso, I swiped open my phone and googled 'Time Machine Memories'. In the video to 'Sorry' there was Peura. Dressed in black and holding his guitar, he stood in the back keeping everyone in time. Better for my analogy if he'd been a drummer, I thought. And, then, is this not the world we want to live in, where our teachers are experts in their craft, committed to doing ever better and setting an example in pursuing their passions.

'We should find the things the student is interested in,' he'd said, 'then invest in that.'

Experts are Made, Not Born

Pekka Peura's practice was born of a scientific inquiry into improving practice that was still rare in education. In 1992, a Stanford economist, Eric Hanushek, took up Bloom's baton with a study of kids in the American Midwest. He wanted to investigate his suspicion that schools could have a much bigger influence on kids' life chances than we realised. After crunching his data, he arrived at a myth-busting conclusion. Schools didn't have a huge effect on kids' achievement. But their *teachers* did. Hanushek saw that, in a single year, kids in the classroom of a great teacher ranked in the top 5 per cent made one more year of progress than they did in the classroom of a teacher in the bottom 5 per cent. The cumulative effect was wild. Students with a high-flying teacher ranked in the top half of the top quartile moved up eight percentile rankings in a single school year compared to other kids, so if you started the school year at 50 out of 100, you ended up forty-second.[14]

If you were lucky enough to have that teacher for five years straight, you'd move up 40 places to the top ten. I'd seen that at KSA.

In Hanushek's view, you could really only judge a teacher's effectiveness in that way. How much did her students learn? You might think it seems tautological – as obvious as the fact that doctors should wash their hands – but that's not always how we've thought about teaching. A bald, bearded, hoop-earringed education professor named Dylan Wiliam has his own take on this. The measure of success in teaching, he says, has long been akin to a doctor stepping back after hours of exhausting surgery to exclaim, 'the operation was a complete success, but the patient died'.[15]

But it didn't stop there. In a separate study, Hanushek made a stark discovery. After the third year of their career, teachers didn't get any better at their jobs. Not if you judged them on how much kids in their classes learned in a year.[16] When doing better mattered so much, it seemed strange that in a profession so tough you could have reached your peak after just three years. Maybe some teachers were just born better than others? Or you might say we've been training them wrong this whole time. A report in the US in 2015 found that despite tens of thousands of dollars being spent on the training of every single teacher, every single year, there was no evidence that any of the interventions helped teachers get better. It was entitled *The Mirage*.[17] The point, as Wiliam neatly put it, was that 'every teacher needs to improve, not because they are not good enough, but because they *can*'.

He believed that the flaw in teacher training to date was that it was focused too much on knowledge and too little on practice.

Teaching was a performance profession, yet trainees primarily learned about it through books and lectures. Traditional training in the US and UK was akin to a doctor spending seven years mostly reading about operations, then walking into surgery alone and being told to get on with it. Changing what teachers do is more important than changing what teachers *know*. Though subject knowledge *is* vital, teaching is also about identity formation, social cohesion and getting the best out of kids. You're a psychologist, a subject expert, a surrogate parent, a friend. It wasn't that we should take knowledge away. But we had to *add* practice – the art of doing better. At Relay, they'd turned to the godfather of it, Anders Ericsson.

Practice Makes Perfect

Ten 20-something trainees lounged in a vast bright room on the third floor of the Relay building in downtown Newark, trading stories from the chalk-face and mulling on the merits of lunch box rice. In addition to the daily class they taught at school, first-year Residents had a rigorous regime of in-class observations, online tutorials and day-long monthly content and pedagogy classes to attend. The pinnacle though was a weekly four-hour afternoon slot devoted to deliberate practice.

'We'll start in ninety seconds,' Dean V. announced over Rock City's 'Locked Away'. *Would you still love me the same?*

The handful of Residents – usually there'd be 35 – moved quickly to their seats. This was remediation for those that had failed gateway three. After several attempts and lots of support, if they couldn't master the basics of 'Introduction to New Material' and 'Engagement' after this session they'd flunk out of the programme after just five months. Expectations were *high*.

'When we're in scrimmage and you guys receive feedback,' began an assistant professor, 'we tend to sometimes feel like we're not doing something right.' She wanted them to shed this feeling. It was vital they rolled with 'in-the-moment feedback' and 'immediate-interrupts' if they were going to really *excel*. She led them in some quick confidence-boosting affirmations, modelling 'Do It Again' when they weren't in sync.

'I can and I will grow,' they intoned as one.

Her style was Da'jia Cornick-sharp. The trainers practised what they preached. Though it seemed severe – and a little controlling – I was beginning to see how the technique served the development of highly skilled teachers. Relay had developed a blueprint for great teaching – and a formula for great teacher training.

Verrilli took to the stage. He thanked them for being there, and for their 'commitment to continuously growing your technique', then took a precise minute to have them play back the fundamentals of the gateway. The pace was high, the cognitive load all on them.

'We're gonna do a quick drill practice.'

The skills would be 'Motivational Wait Time', 'Academic Wait Time' and 'Cold Call', all tools to boost engagement. The Residents had five minutes to prepare their prompts and memorise techniques, then 90 seconds to try them out on their classmates. Drilling built 'muscle memory'. The aim with the starter skills was to achieve *automaticity*.

'The way to get there is practice,' Verrilli had told me on the drive over. Once you knew what you were going for – and Lemov had helped with that – you could get better at it. 'Practising the way an athlete or a musician or a great debater would before they went into the arena,' he explained.

I glanced over at Caleb, Shamira and Deja eagerly scribbling prompts on their pads, mouthing the words, rehearsing their moves.

Game time, I thought.

Relay was applying the insights of Anders Ericsson by the letter. They'd broken teaching down into a set of techniques and were having their trainees use deliberate practice to master these through carefully contrived rehearsal and repeated drilling. The aim was to achieve automaticity in the simpler actions, freeing up vital space in working memory for the trainees to focus on the more important elements of their profession, like what was happening in kids' heads.

Verrilli was the expert-level coach.

'What's the first thing you're going to do?' he asked, pausing intentionally, counting a beat, and carrying on.

'Cold Call, Caleb,' he continued, making his techniques explicit for the fresh-faced young teachers. The quick-fire questions reminded me of Ms Cornick's sixth-grade maths class.

'Reset the class,' said Caleb.

Verrilli 'rebounded' it on to Emmanuel, who built on Caleb's response. Then he changed it up.

'How much time, Lindsay, are we gonna have?' he called out. 'Wait, what did I just do wrong there?'

'The name,' called Emmanuel.

'Right, I didn't time it in. Let me try it again.'

'How much time are we gonna have,' he paused, counted one. One and a half. 'Deja?'

'A minute and a half,' she replied.

'Great, I switched it up and you were ready.'

He was setting a fast pace, modelling engaging classroom delivery, positive framing. It was a pre-game speech by the head coach, but faster, more purposeful. He flipped up an image on the projector, of America's greatest living basketball player, LeBron James. 'All. In.' Janelle rose to practise for the group. She'd have 90 seconds to demonstrate the three core skills live. Two other Residents played the role of kids. After that she'd receive brief feedback, which she'd then have to put into practice. She set herself, looking nervous, then addressed the two Residents playing the role of kids. Drill time.

'Freeze on the opener,' called out Verrilli almost immediately from the back. She'd been walking while giving instructions. Doug Lemov's old mistake.

'Freeze on question, move on narration,' he added.

Janelle restarted.

I'd once watched a film of Wayne McGregor choreographing a ballet. Dean V. danced around the Residents with the same energy and poise. He was *precise*.

'You just softened. Stay confident.'

Janelle was 20 seconds into her 90-second practice. He wanted her to rewind ten, then go again. Get it right this time. Build that muscle memory.

Before heading out to Newark, I'd caught up with Brent Maddin, the Provost of Relay, in New York. He oversaw the design and content of the programme. They offered two main courses. The Residency ran to two years, and was intended as a gradual on-ramp into the profession for graduates of other disciplines. By year two, Residents would have a full-time job, and would work part-time towards a Master's. They also offered a two-year practice-based MAT (Master of the Art of Teaching) to current teachers.

'There's a huge focus on practice,' said Maddin, 'so practise for future performance.' We were sitting in a Midtown coffee shop sipping high-grade lattes discussing why Relay might work where other teacher-training programmes had struggled. Jazz tootled in the background.

'We're closing the "got it"–"do it" gap better than most,' he went on. One of the biggest flaws in the system as it stood was that you could 'get it' intellectually, while still doing it no better as a practitioner (like attempting to learn to teach by reading this book). The usual learning was theoretical and front-loaded. Cast in that light, it was less surprising that teachers tended to plateau after two or three years in the classroom. I thought back to my own chastening teaching experience. Why *had* the training stopped after a year on the job? At its most expert, teaching was at least as complex as maths, music and even medicine. You were a coach, psychologist, community organiser, subject expert, parent for 30 kids at once. But we seemed to prefer throwing our hands up in exasperation – it's complicated! – rather than grappling to define and purposefully grow that expertise.

Maddin puffed out air. He got it. If you only listed the topics teachers had to be competent in – his eyes widened a little in awe – 'then just assigned some very small number to that list, like a day or two, you suddenly realise that there's just *no freaking way* that a novice – or anyone – could do that'.

For that reason, Relay had simplified the equation. Practice made perfect for their trainees, with around 10 per cent of the starter curriculum focused on mastering the classroom management approaches and procedures which they felt most 'helpful and empowering to first-year teachers', and the rest given to the 'teaching cycle' – planning, delivering content and adjusting instruction in real-time. It wasn't intended as a blueprint for super-teaching, but rather a first step on that path. They also focused closely on 'the academic bit', obsessing over their ability to generate and analyse data that indicated with 'relatively small error bars the academic achievement of the students who our teachers are lucky enough to serve'. Test scores were a core criteria Maddin used for evaluating teachers – and he'd studied hundreds of them.

Maddin believed that to develop genuinely great teachers, we should think of a ten-year development window that stretched either side of that first day in the classroom, ending up in a specialism. He referenced a doctor friend now in paediatric internal medicine. 'He specialised!' Expertise was the aim and he defined it in three broad areas: curriculum designer, mentor or coach to kids, and data analyst.

'It would be hard to be an expert in any one of those three,' he added. 'We're asking for a teacher to be an expert in all of them.' You could only become expert if you defined the necessary expertise.

This fitted closely with Ericsson's findings. A raft of subsequent studies had been both heartening, showing that anyone could excel at almost anything, and challenging: it was incredibly tough to do so, and you typically had to start early. Ericsson now classified expertise according to three criteria: the performance had to consistently exceed that of the peer group; it had to result in tangible effect; it should be measurable under lab conditions. In the *Harvard Business Review* he stood by the claim that 'if you cannot measure it, you cannot improve it'. Further, Ericsson could hardly stress enough just how difficult deliberate practice was. It entailed 'considerable, specific, and sustained efforts to do something you *can't* do well – or even at all'. He warned that it was 'neither for the faint of heart nor for the impatient'. Although anyone had it within them to become expert, the process required 'struggle, sacrifice and honest, often painful self-assessment'.[18]

You started with imitation, then graduated to an ability to reproduce from instruction. The next stage was to reliably execute a skill, adapt it to a range of contexts and finally reach automaticity, where you could use the skill to a new purpose. Drills – like the one Verrilli was pushing the Residents through – were crucial.

Relay wanted to get to this level of detail with their teacher training. Where footballers drilled one-touch passing or surgeons practised with their scalpels, so could teachers practise skills. These ranged from the 'super-technical' – 'like the way to time a name when asking a question' – to the far less technical, 'like how you're able to build really deep trusting relationships with an individual kid and her family'. If these things could become automatic, it would open up space to focus on more complex areas of learning. 'If you're actually practising, you are going to be better. Your implementation or performance is going to be better. There's no substitute for that.'

Relay believed that teachers could be *experts*, at least in a big proportion of their practice. And if teachers could be experts, then that expertise could be grown. They watched their teachers every week, videotaped them, put them through stringent, high-intensity practice sessions.

'People have a *ton* of time in clinical placement,' said Maddin. But he also sounded a note of caution. He saw schools that were getting 'tremendous academic results with kids because they are working from a scripted curriculum and they're delivering it with fidelity'. But although the kids were learning lots academically, 'the teachers are leaving', he said. 'Often.' It was reminiscent of the South Korean teacher video. It couldn't be about reducing the profession to a set of moves. This was all about adding skill to the complex work of raising the next generation. When Maddin judged a teacher's effectiveness he looked first at their test scores, but more importantly he asked, 'Would I want to place my child in this classroom?' Maybe there was a little magic after all. Passion mattered. It counted if you cared.

'I want to feel that urgency we worked on,' said Dean V. from the back of the faculty room in Newark. He paced behind the seated students, hand tugging at his chin as the clock ticked round to 4.30 p.m. Scrimmage was under way.

Caleb stood up in front of the group. He looked nervous as he flipped up his opening slide. He turned to face the class, clapping three times in rhythm. The trainees-as-kids clapped in response.

'I need all feet flat, sitting up straight, voice off and hands on your desk,' he began purposefully. Things were going well.

'Thank you, Sam, your hands are on your desk. Thank you.' He paused, forgetting the student's name. 'Stephanie! She's sitting up straight. Thank you for those eyes, appreciate it.'

He walked across the room to stand by Emmanuel, who looked like he was about to get up to no good.

'So there are two common ways that we can identify the way a story is being told. The first one is first person, the second one is third person.' He turned back to the class for a call and response.

Emmanuel stood up.

'Emmanuel, have a seat please! Have a seat!' he said firmly.

Emmanuel carried on doing his thing.

'Take a step towards him, say it again,' interjected Dean V.

Another mock-student laughed at Emmanuel.

Caleb clapped to reset. A couple of students followed him, but the others were gripped by the unravelling chaos. We were one minute in. The thousand impossible moments that add up to a teacher's day came flooding back to me. The road to expertise was long and hard and littered with mistakes. Better to make them here, I thought, where you could stop, think, have Dean V. coach you through a tough moment, than on your own in the classroom in front of students. The rules and techniques still seemed a touch stringent. But these were novices. And didn't you have to learn to hold a pencil, form letters, master spelling and syntax, and practise sentences before you could write? 'Picasso was an awesome realistic artist before cubism,' said Verrilli. 'You have to learn all the rules before you can break them.'

Seeing is Believing

In the afternoon drizzle, Paju English Village looked even more like Oxford, the city it had been built to imitate. Situated an hour out of Seoul, within view of North Korea on a clear day, it was a neatly rendered collection of faux-limestone buildings complete with domes and spires. The village had been constructed a decade earlier as part of an initiative to improve the quality of spoken English among Koreans. Kids would come to Paju for weeks of full cultural immersion, saving them flying halfway across the world for the real thing. Street signs and directions were written in English, in the black wrought-iron style familiar to the old university town. There was even a red

phone box. Apart from the lack of students on bicycles and the presence of Korean-brand cars in the driveways, the one false note was
supplied by a full-size replica of Stonehenge in the car park.

I'd taken a bus out to Paju (against all odds getting on the right
one from Seoul bus station), to visit Chanpil Jung of the Future Class
Network.* Jung was a celebrated documentary-maker at KBC, the
peninsula's equivalent of the BBC. He'd made the programmes I'd
watched about the miserable teachers and sleepy students. His network was now working with Kim Su-ae and thousands of other
Korean teachers to bring about a revolution in their practice. Having
made films about social issues for decades, he had come to think,
just like Oli de Botton, that the only real path to change lay in a
country's classrooms. He'd seen the misery of students and the rise
of the after-school tutor centres that plagued Korean childhood, and
having lost faith in the system decided to take matters into his own
hands.

A few years earlier an unheralded professor at the University of
Melbourne had a bit of a Martin Luther moment, nailing to the door
of global learning his own 95 Theses. *Visible Learning: A Synthesis
of over 800 Meta-Analyses Relating to Achievement* was the most
comprehensive study ever made of kids' learning.[19] For his analysis
of analyses John Hattie had trawled through more than 60,000
studies of student achievement spanning the Eighties, Nineties and
early Noughties, featuring 250 million kids, to rank the impact of
150 different factors from homework to class size. Among policymakers, he became a sensation. His message could not have been
clearer. Learning? *It's all about the teacher.*

The usual suspects had been nowhere to be seen. Facilities. Class
sizes. Uniform. School models. Homework. Extracurricular activities.
Hattie showed that none – not *one* – of these beloved hobby-horses
had any effect on learning. Instead the things that mattered – kids
were being trained to *think*, received clear feedback on how to make
progress, were in schools where adults believed they could learn – all
lay in the hands of teachers. Just as we'd been distracted from the
important stuff in school – the learning – by the things that *looked*

---

* I'll let Grayson Perry know to update his analogy.

important, like uniform, so sight played a role here. The factors at the top of Hattie's ranking were the ones that *made learning visible* to the teachers and, more importantly, the kids. Hattie's message was clear. Learning had everything to do with the expertise and passion of teachers, and very little to do with anything else. Further, if you could perfect Hattie's top three techniques, kids could make an extra year's progress annually compared to the average.

Hattie proposed a moratorium on the usual debates. There was only one thing worth fretting over. How to help *all* teachers become great at their jobs.

Chanpil Jung's solution had been deceptively simple. He'd told Su-ae that in America there was a new craze for the 'flipped classroom' and that *all* the best teachers in the world were doing it. The idea was similar to Pekka Peura's. Instead of standing at the front lecturing the students, you prepared the content in advance, either through recording yourself and uploading a lecture to YouTube, or by preparing resources for students to work through in their own time. The classroom was flipped because instead of doing the difficult thinking and exercises at home, students did it in the class. The teacher didn't just stand and read from the textbook – in that case, what's the point of the teacher anyway? – but gave the lectures for homework, and got students working together in classrooms. Su-ae had agreed to flip her class.

That week I'd visited the classroom of one of the Future Class Network's leading teachers, Gwangho Kim, over in Songdo Future City. I'd watched throughout an entire 90-minute biology class as Kim had sat relaxed at his desk, or quietly strolled the class giving encouragement, prompting new thinking and answering different students' specific questions, as the high schoolers – smoothly dressed in navy-blue baseball jackets with white sleeves – taught each other. The challenge he'd given them was to work out how to conduct a DNA test at a crime scene. First, they had to come up with a methodology, collecting samples, isolating the DNA, working out how to identify its particular makeup, and then compare it to that of a database of suspects. Then they had to design the experiment that would reveal the DNA's code. It was the single most compelling classroom I'd seen.

Working in groups of four, the high school students had found the sources they needed in textbooks and online, and were busy co-designing their experiments. It was electric. Talking to the students afterwards, they revealed to me that at first they were suspicious. This was Korea. The end of school exam – the fearsome Suneung – dominated their every waking thought. They wanted a reliable teacher who would give them a high-grade lecture and put them on a path to acing their test. That's what learning was. But they'd gone with it. The independence felt good. They'd enjoyed being given the responsibility. Even better, the lessons were no longer boring. Preparing for the test was vital, but it was viciously dull. And they noticed something else.

'I realised that with the flipped classroom, we actually learn *more*,' said one student, who was hoping to go on to study medicine.

'I thought it was slow, but now I realise it is faster,' said another.

Gwangho Kim was happy too. He looked relaxed and said that he felt more like an expert. Anyone could read from a textbook and grade a test. Designing learning experiences in which kids took control required real expertise.

There was no single way of being a great teacher, but there were a common set of building blocks. I'd seen Pekka Peura at work, Da'jia Cornick learning her craft. I thought of Ms Toworfe, Mr Jahans and Mr Higgins at Walworth. Each of them defined the role slightly differently, had a particular aim with their work. And yet all of them rooted their approach in a science of improved performance. In their own way, all of them were like Ignaz Semmelweis, pushing learning on from an art to a craft, obsessing over doing better. It was their *métier*, the medium of their creativity, the practice they sought to master. Malala Yousafzai had written that 'if you want to end the war then instead of sending guns, send books. Instead of sending tanks, send pens. Instead of sending soldiers, send teachers.' Teachers are the custodians of our future. We need to raise them onto a pedestal, treat them like scientists, train them like athletes, love them like family.

In part three of my journey, I ask what future we hoped they would build for us. Taking care meant using our natural aptitude for learning and our ability to master the tools of today for good.

I thought we could use education to build a more co-operative and equal world, but I worried that our schools were actually pitting our kids against one another in a dog-eat-dog race for top exam grades. It was true of the UK and US – I'd seen it in Walworth just as I'd seen it in Relay – and it was especially true of Korea.

For now, it was the day of the test. Su-ae had been trying out her flipped classroom for a term, preparing the lectures for students and having them work in groups in class. She looked happy. The kids were smiling and laughing, her classroom full of life. None of them slept. As she projected the results of the kids' latest test onto the whiteboard, there were gasps from the group. Many of the scores had jumped up far beyond their starting point. If we were to stop the flood of teachers leaving the profession all over the world, put them back in their rightful role at the head of schools, then it had to start with pride – and joy. There wasn't yet a formula to teaching, but there was a clear purpose. This was it.

PART III

TAKING
CARE

CHAPTER 7

Big Data

The Examined Life

> If you want to get laid, go to college.
> If you want an education, go to the library.
>
> Frank Zappa

Testing Times

On a late November Thursday in Songdo Future City, Seung-Bin Lee was roused from his trance by the shaking of his hand. He'd been meditating at his desk for 30 minutes, reciting motivational slogans, getting in the zone. He could *do* this. The warm, bright classroom was new to him but the situation was familiar. He'd visualised it repeatedly. During school. At the cramming centre. Even in his dreams. Pencils and erasers were neatly arranged. Energy bars, nuts and chocolate primed to inject timely calories. The trusted water bottle was full (dehydration could be fatal). It was 8.33 a.m. Seven minutes to go. That night he'd lain awake until 2 a.m. telling himself *I. Want. To. Sleep.* But he could not. When his alarm sounded at 5.30, he'd felt surprisingly good. After putting on the loose-fitting T-shirt and comfy sweatpants chosen carefully for the day, he'd made his way to the canteen.

The day was unseasonably hot. Along with 500,000 other zoned-in Korean teenagers, Seung-Bin considered it a possible omen. A year of *fire*. Newspapers carried last-minute tips. Nutritionists proposed high-energy porridge or protein-rich tofu for meals, suggesting recipes for brain-work-boosting snacks. Stylists recommended layering clothes for easy regulation of core temperature, in natural fibres only. Psychologists shared handy suggestions on sustaining concentration. Parents visited temples to pray. But Seung-Bin ignored the annual pantomime. Eating his bland sausage and rice, he had his eyes firmly fixed on his chopsticks, concentrating on chewing each mouthful into optimally digestible morsels. It was monastery quiet in the cafeteria. Kids stared into their breakfast bowls, heads bowed like the condemned. The thought in everyone's minds – *don't choke*.

Arriving at the test centre an hour early, Seung-Bin fastidiously checked the condition of his chair and table before removing his underwear in a toilet cubicle. 'The classroom was hotter than I anticipated,' he explained. Success was in the details, right down to your pants. Invigilators used airport security wands to check for illicit electronic devices and permitted him to use his earplugs. His punctuality ensured tranquillity and he used his extra time to meditate. Everything was as it should be. The parables of oversleeping students forced to beg last-minute rides from police officers to get there on time had done their job. The cops had been readying their motorcycles on the streets earlier. If you missed the start you waited a year to resit. *Unthinkable.*

I thought of Seung-Bin at his desk as I stood outside a test centre near Yongsan station on the other side of Seoul. Well-wishers held banners and cheered the kids as they entered. This was *big*. The stock exchange wouldn't open until 10 a.m. to keep workers out of the kids' way. Planes in the country would be grounded for the 45 minutes of the English listening test. Birds had been asked not to sing. Everything was focused on the success of the teenagers walking purposefully into their exam halls, as hushed and intense as astronauts approaching the launchpad. On a coffee shop TV, I watched a reporter interviewing a group of anxious parents as a gentle wind fluttered the leaves from the trees.

Seung-Bin steadily emptied his mind. Ideal performance came in a condition of *flow*. 'If we start to think,' he told me, 'we can't get a high score.'

Then the shaking of his hand had brought him out of his reverie and he'd looked up at the clock.

Sharpening his mind to a point, Seung-Bin watched the minute hand count down. For eight hours he would be an instrument of pure technique. After 12 gruelling years of school, he faced six short multiple-choice papers. Korean. Maths. English. Biology. Physics. History. 120 times A, B, C, D or E. A few rudimentary pencil marks to decide his future. Which university he'd attend. Who he might marry. His wealth. His happiness. His health. Thanks to his meditation, these thoughts were far from his mind. They were of no use during the hours to come.

At 8.40 a.m. Seoul held its breath. Barack Obama had once urged the world to learn from Korean schools, a miracle of twentieth-century learning. In a few decades, an illiterate, resource-poor country had been transformed into a high-tech nation. Observing the city reach exam fever-pitch, I worried he'd been mistaken in his enthusiasm. Korean kids were paying a high price for their success. The university entrance test cast a long shadow over the country.

Seung-Bin steeled himself. His hand ceased shaking. The world's toughest exam was beginning. They called it the Suneung.

What Can't be Counted Still Counts

Data drives the digital era, but we must ask what it's doing to our kids. Though the test Seung-Bin sat was an extreme example, it is part of a global trend that pits our young against one other in a no-holds-barred race to the top. The competition has brought global progress in literacy and numeracy for millions, but it has also left many behind. Throughout my travels I sensed that it came at a cost to young people like Ifrah, Lilas and Seung-Bin, diminishing their well-being and binding them into an ever-tighter straitjacket of success. Finland's classrooms reminded me that we weren't born only to think and to do, but also to care. Yet it frequently seemed that today's kids were so focused on exams that the space for developing their human capabilities was being squeezed out. In the final part of my adventure, *Taking Care*, I hoped to find out what it might mean for us to learn to pursue the fullest expression of our human potential, setting kids up to be happy and to co-operate in shaping a shared future. I'd started in Seoul in an effort to answer a simple question. Was big data the poison, or the cure?

It's been 20,000 years since our ancestors on the banks of Congo's Semliki River first used quartz to score marks on the earliest known tally stick, hinting at the existence of a Stone Age Silicon Valley of ground-breaking maths prodigies and Upper Paleo diets. By the time of the Mesopotamian cradle of civilisation 5,000 years ago, the creative genius of the Ishango data entrepreneurs had been co-opted by the Sumerian state as a tool of administration. The cities of Ur, Lagash, Eridu and Uruk were each led by a priest-king supported by a team of bureaucrats who surveyed and distributed land and crops. Assiduous record keepers, they developed a new form of writing, cuneiform, to keep track of goods and crops on clay tablets, which they stored in great stacks, a forerunner of today's state IT projects, only more reliable.

The trend for government data tracking continued in ancient Egypt, Greece and Rome, before a new idea dawned during the Age of Reason, when we asked whether we might analyse data to better understand ourselves.* This gave us Auguste Comte's sociology and an explosion of utopian ideas about how to organise human society according to the rules of science.** Among the ideas of the age were those of a Belgian astronomer and statistician, Adolphe Quetelet. His particular 'social physics' was built around the idea of *l'homme moyen*, the scientifically average man who stood at the centre of normal distribution on all measures: height, weight, intellect, appearance. Todd Rose reports in the *End of Average* that astronomers

---

* The word Statistik was coined in Germany in the eighteenth century, and means literally 'a science of the state'.

** Auguste Comte is the most prominent of the early socialists who emerged at the start of the nineteenth century. Observing the role of science in organising knowledge and industry in particular – this was the era of the steam engine and industrialisation – the early socialists attempted to apply scientific thinking to the organisation of human societies. The most eccentric of all of these early socialists, and my favourite, was Charles Fourier, who understood the inherent genius and industriousness of kids, envisaged a world in which humans worked in harmonious free-love-practising phalanxes of 900 people, and predicted that when science had finally perfected the operation of the world, the sea would turn to lemonade.

of the era used statistics to study the movement of heavenly bodies, with Quetelet realising that the same models could be used to analyse social factors influencing human behaviours like marriage, crime or suicide. He seized on newly available sets of big data, most famously using a record of the chest sizes of 5,738 Scottish soldiers to propose the first-ever measure of BMI. Quetelet's Average Man became the norm around which we built our schools for an average student progressing at an average pace through an average amount of learning.

If we are to create a truly caring approach to education, we have to challenge this idea. Data promises to oil the mechanisms of our societies, but it also risks reducing our natural-born learning potential to something equally machine-like. In order to understand if data could be a tool for care and co-operation, rather than competition, I turned to another fabled statistician, based not far from Quetelet's birthplace.

Without Data, You're Just Another Person with an Opinion

Barack Obama's praise for South Korea's schools could be traced to a moment in December 2001 when a slender young analyst with broad ambitions sat down with ministers of education in his native Germany to present a report. For five years he'd begged, borrowed and stolen – 'I had a zero euro budget' – from his workstation in Paris, uniting a crack team of stattos to carry out the research he'd just presented. The project had its roots in his university days in Hamburg, where as a subject-hopping physics undergrad he'd been intrigued one morning by the ideas of T. Neville Postlethwaite, a self-styled 'educational scientist', as he lectured on the potential of applying hard-edged thinking to the so-called soft disciplines.[1] A brief intellectual bromance alighted upon a common passion: applying a scientific, data-driven approach to the evaluation of human learning. Its effects would ripple through every classroom in the world.

The report was his magnum opus, unlike anything seen before. The data avengers tested 300,000 15-year-olds across 32 countries, crunched the numbers and framed the findings. Six years earlier he'd attended a meeting with the top policymakers of the world's largest economies. 'Everyone was saying, "We have the best education system in the world!"' he told me. They couldn't *all* be right. His work would decide it. Mindful of his professor father's warning that it was impossible to 'measure what counts in education, the *human qualities*', he'd worked tirelessly to craft a new type of exam that would probe more than memorisation, evaluating the skills really required to thrive as a knowledge worker in the twenty-first century, 'to analyse, compare, contrast, critique', he said, 'to think like a mathematician, to think like a scientist, to think like a historian'.[2]

The results would reveal for the first time whose kids were the smartest in the world.

The German ministers were unmoved. 'People told me, "It's just another one of these studies, people will forget it,"' he said, pausing for effect. 'And then, on the day of the launch, *it was everywhere*.'

Today the publication of the new Program for International Student Assessment is a hotly anticipated global event, taking place every three years. In the latest round, 540,000 15-year-old students in 72 countries were tested on their science, reading, maths and collaborative problem-solving skills. The hotbeds of talent were in East Asia and Scandinavia, with countries like Singapore, China, Korea and Finland leading the way. British 15-year-olds placed some way off Olympic standard at just above average,* while American teens languished in the bottom half of the table. 'On the world stage, U.S. students fall behind,' went the headlines.[3] The tests still had the power to shock.

PISA's mastermind was Andreas Schleicher. On a grey afternoon I journeyed out on the Paris Metro to meet him at the OECD's headquarters on the Quai Alphonse le Gallo. Striding along the riverbank opposite the Bois de Boulogne, the mansions of Old Europe standing solemn sentry over the gleaming glass frontages of the New, I'd sensed how fast the world was changing and how radical it was that we could even compare kids in such far-flung corners. Schleicher was fêted by the *Atlantic* as 'the World's Schoolmaster'

---

* Specifically, we Brits came in at 22nd in reading, 27th in maths and 15th in science, our teenage brainpower falling at roughly the same level as our footballing prowess.

and 'the most influential education expert you've never heard of',[4] his calling to 'Use Data to Build Better Schools'. PISA was his method. Michael Gove, the UK Education Secretary, had called him 'the most important man in English education' and, to Schleicher's particular pleasure, 'the father of more revolutions than any German since Karl Marx'.[5]

'Tradition is the enemy of good practice, really,' he told me, leaning forward in a box-fresh office chair. It was his first day at the new HQ, a sweeping glass and plywood cathedral to the globalising mission of the OECD, the Organisation for Economic Co-operation and Development that was founded to forge closer global ties among developed nations after the Second World War. Still slender, with blue eyes, a white helmet of hair and brown moustache – less Karl Marx, more *Mittel*-European minimalist jazz saxophonist – learning's data-driven revolutionary was in a relaxed mood. 'What PISA has created is a mirror where you can see yourself in light of what others are doing.' Schleicher worshipped at a church of numbers.

If we lived in our own national bubbles we would never see the world in full, he explained. His data was a further step in our enlightenment, bringing a torch to the pagan caves of learning and continuing Auguste Comte's mission. He was much in demand. That morning he'd landed from Santiago and Rio and would depart soon for Abu Dhabi and DC. His genius had been to generate data that allowed the world to compare what worked and what didn't across multiple countries. Rather than pay attention to learning inputs, things like teacher salaries, class sizes and curricula, his data had shifted the global conversation to learning outputs. What did kids in a given country know? What were they able to do? 'We've designed education systems and schools around the people working there and not around the needs of the children,' he explained. Instead we should be asking, 'How well are our students prepared for their life?'

His research was the first to tackle the question. The report was a jolt of shock therapy to national education systems.

'It was taken up by every channel,' he said in his clipped, controlled English, 'and not just for the day, but for *weeks.*'

The German ministers were agog. They'd firmly believed that their schools were the best in the world. As had the Brits, the Americans, all of the major economies. But in PISA 2000 it was Finland – *Finland!* A country of ski-jumpers and Moomins! – that had come out on top. Used to giving press conferences to one woman and her dog, the Finnish education minister was astonished that morning to find the TV cameras of the world's media pointed at her. Meanwhile, German students performed *below* the OECD average in maths, literacy and science. Worse even than the Americans! They had come in at mid-table, below the above-average Brits. Boosted by their outstanding maths and science scores, Japan and Korea completed the top three.

The data was incendiary. A *Der Spiegel* headline demanded 'Are German Students Stupid?' *The Economist* led with 'Dummkopf!' The ministers, flailing around in blind panic, called for Schleicher's resignation.

'At the start, only a minority accepted it,' he said, smiling. Behind him two block-printed images of children at play leaned unwrapped on the filing cabinet, along with photos of some of the conference crowds he had spoken to. 'The majority of countries said it cannot be done, financially, methodologically, logistically. It *shouldn't* be done.' They criticised his approach, suggested the assessments were too limited in scope, lamented a growing obsession with testing. Schleicher was unperturbed. 'We have to accept that some things are very good indicators for many things,' he went on. 'The absence of mathematics skills doesn't imply the presence of social skills.'

People misunderstood and misused data. 'The concept of an indicator,' he sighed, 'it's not the thing. It's just an *image*.' PISA was only 'a mirror'. He understood of course that when countries caught their reflections, they might not like what they saw.

'It's often that gap between who you believe you are and who you actually are that surprises you.'

After the shock, PISA became a household name in Germany, spawning *The PISA Show*, a prime-time TV quiz. Ministers pored over the findings, coming to surprising conclusions. Things believed to make a big difference, like money spent per pupil, class size, curriculum content or sending your kids to private schools, were shown not to matter. Whereas the UK and US had always said income inequality was the reason why poor kids performed worse at school, other countries like Finland and Korea proved poverty didn't have to be destiny. 'The 10 per cent most disadvantaged students in Shanghai outperform the 10 per cent most privileged in the US,' he explained. 'That is what is possible in education.' Data like that *was* revolutionary – he had that in common with Marx – though the statistics predicted a crisis for our kids. 'The gap between what the world needs and what we offer is just becoming wider,' he said. Poor education meant greater class divisions, social alienation and a risk of radicalisation.

His mission was to give governments a gift of insight and the tools with which to do something about it. Toting a PowerPoint deck on his hip, he'd become a global education pin-up. But in the grey afternoon light he might have been a cyborg sent from the future to save our kids. His data saw deep into the soul of our schools. It couldn't lie. Used well, it might fuel human understanding and judgement, enabling us to take better care and fulfil our natural potential. Used badly, it reduced learning to a narrow box-ticking exercise. Although it *was* only an image, I reflected that we often took it for the whole.

He looked down at his watch, a smooth obsidian square nosing out from the red sleeve of his jumper. He shook it, but the batteries were out. Technology could take over the world when it fixed its power sources. Still, we were out of time, and we'd continue our conversation on the train back to Paris. As we gathered our belongings, the words of his father echoed in my head. 'You can't measure what really counts, the human qualities.' There were no global tests for art or music, well-being or relationships. Perhaps there never could be. It was important to pursue great data, but countries had become obsessed with the three-yearly rankings, fuelling a growing exam fever.

The Race to the Top

The top-three ranking of Korea's 15-year-olds at the turn of the millennium was a miracle. Following the Korean War 50 years before, the country had been destitute of both income *and* learning. Hemmed in by the superpowers of China, Japan, the USSR and

USA, its prospects were bleak. GDP was the same as Ghana's and it depended on foreign aid handouts to survive. On top of that, four in five Koreans were illiterate. Asked to back a country to thrive, you'd have bet on Ghana every time. But today it's *Korea* that has the thirteenth-largest economy in the world. In five decades the country's GDP grew more than 40,000 per cent. Samsung, Hyundai and LG became global mega-brands. 'We don't have any resources,' said Ju-Ho Lee, Korea's former Minister of Education, Science and Technology, when we spoke, 'just our minds and hard work.' The miracle was manmade, powered by the country's inhabitants and the result of a multigenerational effort to grow their brainpower.[6]

Today Korea is a *talent* superpower, locked into a virtuous cycle of learning begetting learning. Since the Sixties economic growth has been powered by the development of the country's 'human capital', which in turn has meant more spending on education and more money-making Korean intelligence. These days just 2 per cent of South Koreans are illiterate. Among the 50 million inhabitants, 25 million of whom live in Seoul, are a greater proportion of university graduates per capita than any other country in the world. In short, in half a century to 2010, no nation learned more. And at the pinnacle of this effort was the Suneung, the six-subject, eight-hour multiple-choice exam that would evaluate – and rank – every Korean high school leaver at the end of their education. It had seen their kids rise to the top of the tables, fuelled Korean knowledge industries. But the success came at a price.

At the end of his exam in Songdo Future City, Seung-Bin felt deflated. With five minutes to go in his last paper, biology, he'd completed the questions and had time to reflect on his school career. 'It was pleasant and I regret it too,' he told me over Skype a fortnight later. He was relieved to be finished, but overawed at the 12 long years of effort he'd put into his studies, all for the decisive day of the SAT. While other students had screamed or sobbed at pencils down at five-thirty, he'd gone quietly to the school dormitory to sleep. He could think of nothing else to do. 'I concentrated all day. I was so tired,' he explained. This empty feeling was common to many Korean high school students. The country's economy-boosting investment in learning paid off. Kids were inspired to work hard.

But it had morphed into womb-to-work *Hunger Games* where only the strong survived.

'It is insane,' he said on further reflection. 'I used to have a dream about going to university. After I entered high school and did a lot of tests, I changed the dream to a fear.' The system was meritocratic, but brutally so, with nowhere for students to hide. In the end all that mattered were your test score and national ranking. The French philosopher Michel Foucault described the high-stakes test as an instrument of 'governmentality'. 'It establishes over individuals a visibility through which one differentiates them and judges them,' he wrote. The true aim of exams was not to inspire learning or measure outcomes, but simply 'to qualify, to classify and to punish'.[7] In Korea his dictum was writ large. They didn't disguise it. Hard work was a fetish. Every second counted.

Seung-Bin showed me his school timetable for the three years of high school. Every day began at 7 a.m. with self-study time, before first class at 8.30 a.m. He'd then spend the day in 'boring lectures' until 4 p.m., when he'd returned to self-study until 9 p.m. That wasn't the end of it. After evening self-study time, he would journey over to the *hagwon* – a private tutoring centre – to get his *real* learning done, honing his exam technique, finally finishing at 11 p.m. A 14-hour work day was perfectly normal for the three years of high school. Weekends were considered valuable learning time too. On Saturday and Sunday he'd gone again to the *hagwon* for a relatively restrained 12 hours from 7 a.m. to 7 p.m., albeit 'with self-study in the evening'. When I asked about holidays or time with friends he said, 'I watched a DVD once a month.'

In his final year, work was even more intense. Seung-Bin had emailed over his timetable and it appeared to show him at the *hagwon* from 10 p.m. until 2 a.m. each evening. 'That's right,' he told me. This was despite a police-enforced curfew imposed by the government to shut the centres at 11 p.m. Korean kids managed five and a half hours sleep on average, but most I spoke to claimed three or four. The nation was sleep-deprived. In the race to the top, rest was a luxury you could not afford.

Driving kids like Seung-Bin on was the Suneung. You didn't get into university thanks to cultural capital, debating, arts, sports or having the right accent, but through your ranking, the hundred or so careful pencil marks you scratched into the multiple-choice answer boxes. It meant that the Suneung too wasn't open to interpretation. Your answers were either right or wrong. A bigger concern voiced by Seung-Bin and Ju-Ho Lee was whether the exam tested skills with any real-world applicability. They worried that acing the Suneung no longer aligned to succeeding in today's world. 'We don't have to think, we just compare the paragraphs and the questions,' explained Seung-Bin. 'When we think, then we choose something wrong. We just have to think inside the paragraphs and choose the right answers.' The test seemed to ask kids to become word-processing machines. You learned above all the *technique*.

Consider this question from a Suneung English paper. And set a stop-watch. Candidates had around sixty seconds to read the paragraph and arrive at an answer.

So far as you are wholly concentrated on bringing about a certain result, clearly the quicker and easier it is brought about the better. Your resolve to secure a sufficiency of food for yourself and your family will induce you to spend weary days in tilling the ground and tending livestock; but if Nature provided food and meat in abundance ready for the table, you would thank Nature for sparing you much labor and consider yourself so much the better off. An executed purpose, in short, is a transaction in which the time and energy spent on the execution are balanced against the resulting assets, and the ideal case is one in which _____. Purpose, then, justifies the efforts it exacts only conditionally, by their fruits.

(1) demand exceeds supply, resulting in greater returns
(2) life becomes fruitful with our endless pursuit of dreams
(3) the time and energy are limitless and abundant
(4) Nature does not reward those who do not exert efforts
(5) the former approximates to zero and the latter to infinity[8]

As I toured schools in Seoul, Yumi Jeung, a journalist who'd lived in Finland and now worked for the Future Class Network, teased me with these questions, which I failed repeatedly, protesting a lack of concentration. This pastime had gone viral. Native English speakers ran YouTube channels capturing their feeble attempts at answers. Jeung also pointed out that you could score full marks on the test without ever learning to speak or write a word. The tests were scrupulously fair – the data could only lie to the extent that you could guess answers (Seung-Bin judged himself to have done better than he should have thanks to a couple of lucky guesses) – but also weirdly skewed. In a bid to create the most objective exam possible, the designers of the Suneung had created a strange hybrid that demanded kids think and act like algorithms.

The real reason Korean kids did so well on PISA, according to Ju-Ho Lee, was their indefatigable commitment to hard work and self-study. They succeeded *in spite of* schools. The country had a secret weapon in its *hagwon*s. Parents spent $20bn a year sending their kids to private tutoring centres to get them ahead in the race to the top.[9] At places like TIME education in central Seoul, a parent and child could meet a specialist consultant who'd take a look at their academic record, listen to their hopes for university, diagnose a set of needs and prescribe customised private tutoring sessions for you to reach your goal. Tutors lived or died by their reputation, with some online impresarios earning as much as $4,000,000 dollars annually. Results paid. Sitting in on sessions at TIME I was stunned by the minutiae of Suneung preparation. Tutors picked apart the sentences of test passages like linguists poring over the Dead Sea Scrolls, annotating every word, scrawling reams of hieroglyphics on their whiteboards. It wasn't learning, but *deciphering*, code-breaking. 'You don't start by reading the paragraph!' Jeung had laughed when I failed again at an English question. I'd never been good at cryptic crosswords.*

The exam wasn't only skewing learning. Korea was in the grip

---

* Like all good crossword-setters, the maestros of the Suneung set out to puzzle the young hopefuls. Jeung explained that you began of course by analysing the sense of the answer statements.

of a mental health crisis, with the highest suicide rate among OECD member countries and a surge among its teens.[10] A few years before my visit, the country had been gripped by a gruesome murder case in which an 18-year-old boy was found to have stabbed his mother to death, sealing her body up in a bedroom and living with her remains for eight months in the home they shared in east Seoul. He'd been fabricating his grades since middle school, most recently boosting his position from 4,000th nationwide to 62nd. She'd told him even that wasn't good enough and forced him into ten hours of press-ups and beaten him with a baseball bat. He'd snapped, pushing a knife into her neck. Recognising the relentless pressure of the Suneung, the public had sided with him and he'd been sentenced to just three years in prison. His true grades had placed him in the top 5 per cent of all students.[11]

'Koreans are not happy with it,' said Ju-Ho Lee of a system where your career, health and happiness came down to your data. It was right for an age of economic growth in which you hoped to pull a nation up by its bootstraps, but today it was outdated. 'Test scores may be important in the age of industrialisation,' he told the *Washington Post*, 'but not anymore.'[12] While back in the UK we were introducing more exams at earlier ages, Korea was planning to reform its system around creativity and social and emotional skills.

A month after my visit, I spoke again to Seung-Bin over Skype, his exams now a distant memory. He'd scored ones and twos out of seven in his Suneung (the equivalent to A*s and As at A level), which qualified him for a top-three university. After that he'd taken some time off for the first time in years, learned to drive, had his hair cut into a new style after the icons of K-Pop. He was in a philosophical mood and had prepared me a handy chart of Korean suicide statistics and written an account of the day of his Suneung. 'Every year there are accidents,' he said. He wanted people to understand the pressure South Korean students were under, ground down by the relentless quest to perfect their data.

'If my friends do the study and I can't do the study, that is *most* stressful,' he explained. *Because it was a competition?* I asked.

'Yes, yes! Right!' he said. 'My high school had a lot of activities, group activities. But when we did the group activities, if two people are participating in one project, it is really important to have the

balance between the two guys.' Otherwise, you risked your partner running off to revise while they were stuck with the project.
I was struck by how many of the dozens of conversations I'd had
with Koreans about school felt emotionally charged. A successful
30-year-old entrepreneur broke down in tears recalling the pressure
of her teenage years, telling me that her hair had fallen out. At an
evening meeting with five otherwise unremarkable students in their
early twenties, two revealed they'd attempted suicide as teenagers.
Over dinner they traded school stories like shell-shocked soldiers
returned from war.

On one level, it was hard not to admire the Korean system. The
model had powered a learning miracle, proved the extraordinary
potential of our minds. It *worked*. But was this truly the future we
aspired to for our kids?

'Most stress is from studying,' Seung-Bin explained finally. He
was happy to be free. 'You have lots of things to study, and you have
to soak up the pressure to get high scores.' Ju-Ho Lee was concerned
the Korean bubble was about to burst. The one-size-fits-all work-
till-you-drop dogma of Korea's knowledge workers was wearing
them down. The strange cognitive demands of the Suneung seemed
destined to prime kids for automation. Korea already had the highest ratio of robots to workers in its manufacturing industries,[13] while
NAO robot teachers were supposedly being used in some places
to instruct English. I wondered how Seung-Bin had dealt with the
stress. Despite working 15 hours a day, seven days a week for the
last year, he and his friends could only think of one solution. 'It's
kind of a crazy answer,' he said, 'but I think we have to study *more*.'

You could see it as a virtue. But it was also a symptom of an
uncaring system that denied children a childhood, restricting access
to the white spaces of freedom where we practised emotion, learned
empathy, developed our well-being. And I knew from my own classroom just how widely this logic had spread.

We are All Scientists Now

By 2008 the use of data had filtered down to Walworth Academy.
'Here it is,' Mr Saunders told me on the first day back after the

summer, handing me a slim blue file. Tanned teachers spoke of Greek islands and a Nescafé aroma warmed the air. I'd not been entrusted with the life-or-death stakes of the Year 11s – I was newly qualified – but would be teaching classes 7X1, 8Y2, 9X3, 10X4 and 10Y1. I opened the file. Inside, a couple of stapled sheets held the data. On top were student names, a previous teacher comment and predicted GCSE grade. Below, an eye-watering spreadsheet with a shifting mosaic of numbers and letters that outlined each pupil's on-track end-of-year grade, current working grade, and the results of every key assessment they had taken since the end of primary. It was expected that pupils would make two sub-levels of progress in every year they were at school. Their current progress data was helpfully colour-coded in a dot-matrix of traffic light colours: can-do-this green, slowly-getting-it yellow or not-a-clue red. I understood my mission for the year. All greens.

Our schools had no choice but to embrace the use of data. Learning was a vast and complex field, with 9 million children in school in the UK alone, each one an individual specifically shaped by their genes and experience, uniquely able in some areas, uniquely challenged in others. If you were Aristotle tutoring Alexander the Great, you could tease out an individual pupil's abilities, as Bloom had shown. If you were Alex the Not-So-Great-Is-Putting-It-Mildly struggling to teach 150 kids at an inner-city school, you needed a shorthand for what each kid knows and can do, in the form of grades, reports, work samples, test scores. We just had to ensure we didn't start to take the shorthand – 'the image', as Schleicher put it – for the whole. They were three-dimensional human kids. Data only told us so much.

The numbers had a powerful logic, however. That year we tracked kids' progress against specific National Curriculum objectives, carefully aligning our practice to a set of learning outputs. Though each teacher cared deeply about individual kids, the logic of the approach focused us on grades. My two groups of Year 10s with huge gaps in their reading and writing abilities all *had* to achieve a C. Under pressure, I looked for a short, easy text to study and chose just a few sections for us to read. I carefully analysed the ingredients of a GSCE C-grade essay so that they could be replicated. We'd not learn *all* of the Sherlock Holmes story, *A Study in Scarlet*, but we'd take the most efficient route to writing *one* successful essay about it. If each pupil answered the same question, with broadly the same examples, arguments and answers, we'd save time. It was less a learning science, more exams by numbers.

At the time, I was pleased with my efficiencies after the horrors of my first year. The coursework essays were a success. GCSE examiners paid no attention to whether the kids had read the whole book, or developed original opinions about it. I hadn't become a teacher to game the system or maximise the productivity of my classroom, but that's how it had ended up. Sometimes you found yourself teaching to the test, so that when they took the assessment it *appeared* like they'd learned a lot. But, on reflection, when my kids got their grades, I was left with that gnawing sense of what might have been. The use of productivity data had fuelled growth in our twentieth-century economies, but it was also ultimately responsible for sweatshops and zero-hours contracts. At worst, if our schools became *too* stuck on data, we risked creating systems that forgot we were human.

The highest-performing in the world could be found in Shanghai. Their kids had topped the 2012 PISA charts. I'd flown on from Seoul to find out their secret.

The Smartest Kids in the World

In a salmon-pink six-storey municipal building amid the high-rise apartments and swooshing electric vehicles of central Shanghai, 36 third- and fourth-graders were arrayed in tightly packed boy–girl, boy–girl rows through a fourth-floor classroom at Wanhangdu Road Elementary School. Below us on a green exercise track a phalanx of younger kids were comically soft-pedalling through the press-ups, star-jumps and jogging the PE teacher was demanding of them. Inside, the kids sat keenly awaiting their teacher's instructions in neat white polo shirts and red neckerchiefs like miniature sea captains were heirs to the future. The city was booming and so were its kids. For the past six years Shanghai 15-year-olds had topped the global learning charts.[14]

In the lobby downstairs a hand-painted mural of a moon landing urged: *Come to participate! March to future!*

At 9.15 a.m. a jingle chimed through the school and Selena, ten, ran from her seat to the blackboard, turning to face the class. As music piped from a speaker high up on the classroom wall, she began to sing.

'My school will shine today,' warbled Selena. Her classmates leapt to attention at their desks and chorused, 'My school will shine.'

It was feel-good, all these kids in uniformed unison. In my head, I joined them. *My school will shine today, my school will shine.* I knew how important rituals were to children. And yet, was it a *little* like brainwashing? Dismissing their top scores, critics accused Shanghai schools of mass-producing unthinking high-performance drones that aced tests but lacked social and emotional skills to succeed in the world. I'd come here to find out the truth. If the tech-utopians were reinventing learning, the kids of East Asia were the ones acing tests. Shanghai was providing the best basic education in the world, winning the race to the top.

'My school will shine today, my school will shine,' they indoctri-chanted through three more rounds. I'd got my first sense of the Shanghai approach from *Are Our Kids Tough Enough?*, a UK TV show that flew two Chinese maths teachers to a Hampshire secondary school to compete against their British counterparts. Despite not knowing the kids or context, they'd triumphed when the two groups of pupils were tested at the end of the series. I was mindful that Schleicher's PISA tests were designed to measure more than rote memorisation. They were a test of *thinking*.

The happy sailors moved into a brief office gym routine at their desks – stretch, shake it out, deep breaths – then sang again in Chinese. By 9.20 a.m. they were ready, taking their seats in boy–girl pairs at the three rows of six tables. Ms Jingwei, a young, newly qualified elementary school maths teacher in a woollen poncho and stylish large spectacles, took up her position by the projector. I clutched an English translation of the lesson plan. How could we use number lines to express fractions?

It wasn't so surprising that kids in Shanghai excelled, or that it was dismissed as an exam factory. China had invented the race to the top, the world's first standardised tests given 2,000 years ago by the Han Dynasty. Aspiring civil servants had been made to sweat through the *keju*, a test on military strategy and Confucian classics. For three days candidates were locked in purpose-built cells at the imperial academy, eating and sleeping in cheat-proof cubicles as they scribed answers in essay form. The practice was spotted by officers of the East India Company at the time of the Qing dynasty and introduced for entry to the Indian Civil Service in the mid-1800s. Within a few years, the UK's first standardised school-leaving exams were being put out by Cambridge University. Alongside mass education, the standardised exam became ubiquitous. My 10X4 class had Emperor Wu to thank for their coursework travails.

The day before, I'd gone over this story with an ed-tech entrepreneur, Stella – her latest AI could robo-grade and auto-feedback on up to 15 sentences of writing and would soon do the same for speech – and her friend Zhang Mingsheng, the retired Deputy Secretary-General of the city's Education Commission. As we ate lunch in a dark wood-panelled dining room at the Phénix, a French restaurant overlooking the fir trees of Jing'an Temple Park and skyscrapers beyond, Zhang told us an ancient poem.

'In the morning, we want to work with cows and sheep,' he said. 'In the afternoon, we want to work for the Emperor.'

I nodded sagely.

Learning had been in the fabric of Chinese culture since Confucius. A hundred years before even Plato's birth, he was teaching that if your plan is for one year then plant rice, if your plan is for ten years plant trees, if your plan is for a hundred years *educate children*.[15] He was revered and his rote-learning methods continued to influence Shanghai's approach, with kids arriving at understanding through repetition and memorisation. I'd seen it in Ms Jingwei's maths class. Following the song, three students quickly stood, explained in turn where you would find a half, a third and a quarter on a number line and sat back down. Kids then worked independently for five minutes, mapping 3/10, 3/5, 1/7, 1/9, 3/7, 9/10, 7/9, 5/9 onto four different number lines. Again, Ms Jingwei asked a few of them to explain. Next came classification of fractions. Was 3/5 or 3/7 larger? What about 7/9 or 5/9? Was 9/10 greater than 3/10? 1/9 more or less than 1/7? Kids had a quick go and reported back, the number lines helping them to visualise the differences. She then opened a discussion. What were some other ways of comparing fractions? One girl suggested multiplying them so that the denominators were the same size. 1/9 and 5/7 became 7/63 and 45/63. Then it was *obvious* which was bigger.

From my seat at the back, I could see the layers of learning building up. Repetition led to mastery, as Zhang had explained. You took a concept, in this case fractions, and explored it from multiple angles – number lines, multiplications, greater than or less than – to ensure mastery. A game with cotton buds brought the lesson to a close after 35 minutes. I could almost sense the cognitive architecture laid down in their brains.

'Clever students!' exclaimed the translator as the kids assessed themselves on scraps of paper. They were to self-evaluate, as in Pekka Peura's classroom, on behaviours like learning intent, learning habit and learning efficiency. The lesson was precision engineered. Chinese teachers were typically subject-area and grade-level experts, teaching the same lesson repeatedly, refining the particular craft, say, of teaching fractions to eight-year-olds. This made for efficiencies on the teaching side, but needed all kids to move at one pace through the system. There were extra benefits. Schleicher pointed out that Shanghai's lowest performers were scoring higher than the top performers in the US. But not everyone was a fan.

Yong Zhao, a Chinese education expert based in the US and author of *Who's Afraid of the Big Bad Dragon: Why China Has the Best (and Worst) Education System in the World*, explained that Shanghai teachers and parents felt like they were caught in a trap.[16] They wanted a multidimensional twenty-first-century education for their kids, but couldn't escape the current logic of the race. It was a Prisoner's Dilemma. If everyone opted out of the race, fine. But if even a few people continued to compete – getting names down for top nurseries, investing in tutoring, honing test technique – then, rationally, you had to stick your kid in the starting blocks as well. No one wanted to risk losing out. But as the Nash Equilibrium, the subject of *A Beautiful Mind*, had shown, the best outcome for the individual didn't necessarily mean the best all round for the group.* The race could become pathological. China's Gaokao exam provoked the same mass hysteria as Korea's Suneung. Top courses at elite universities had been known only to accept one in every 50,000 candidates.

Zhao argued that although PISA scores said kids in Shanghai could *think*, at best the authoritarian imperial tradition developed a narrow type of intelligence. At worst it meant a rigid culture of competition and compliance with everyone desperate to grasp a few coveted government jobs. 'China,' he wrote, 'has produced the world's best test scores at the cost of diverse, creative, and innovative talents.'[17] Confucius stood for learning. But he also embodied imperial authority. Shanghai's education system was highly effective in Zhao's view. But it was also on some level an authoritarian machine inputting into the head of every kid the precise content the government wanted them to learn. Qualify, classify, punish, said Foucault.

We couldn't dismiss the achievements of Shanghai or Korea – Schleicher had told me that 'the system is good at attracting average people and getting enormous productivity out of them'. The students at Wanhangdu Road Primary School and Wucai High were full of life. Their teachers were smart, proud professionals, the envy of the world, as the Hampshire comprehensive schoolkids discovered in *Are Our Kids Tough Enough?*. And yet Zhang stressed that Chinese policymakers were as interested in studying the UK and US approaches as we were to learn from Shanghai. (And we were *keen*. In Wanhangdu Road I bumped into four maths teachers from Burnley, over as part of a larger group on a Department for Education-sponsored fact-finding visit.) Yong Zhao explained that

---

* In the Prisoner's Dilemma, two suspects in separate cells are asked to inform on one another about a crime. The question is whether criminals A and B should co-operate or defect. If A remains silent and so does B, both will get a year in prison. If A informs on B, who remains silent, A will get off and B will get three years, and vice versa. If both A and B inform on each other, they'll each get two years. The dilemma is: what should A and B say? The Nash Equilibrium explores the idea that although the rationally self-interested option would be to inform on one another, paradoxically *if* the two of them co-operate, they will get a better outcome (a year inside each, adding up to two in total). If you watch Russell Crowe carefully in the movie version of Nash's life, you can clearly see him thinking all of this.

China's economy had dominated world output until the eighteenth century, when the European *thinking* revolution broke the trend. The scientific, liberal-humanist model of the Western powers had led to a learning and technology advantage that saw the great empire subjugated. Today in China, as in Korea, the question was how to instil greater creativity and critical thinking in kids, without also disrupting the hierarchical ties that bound society.

Lucy Crehan, a British teacher and author, whose brilliant book *Clever Lands* tours the world's leading education systems, recounted a Shanghai head teacher likening Chinese schools to Kentucky Fried Chicken. 'They don't have amazing chefs in every store,' he said, 'but they have certain procedures in place.' The city had adopted an educational Taylorism, narrowing the focus, standardising content and procedures, and refining the processes. It worked. Shanghai's kids were great readers and writers, mathematicians, scientists. But it felt like a narrow definition of learning in today's fast-changing world. Shanghai succeeded because it had created the ideal *twentieth*-century system. Every second kids spent in class went on mastering content, learning thinking as an instrumental tool. But the focus on mastery squeezed the system. Tight control of the environment meant kids lacked freedom. From the roots of Shanghai's success grew the shoots of future challenges. No one could afford to blink first.

At Phénix the espresso and French pastries reflected Shanghai's internationalism. Like Andreas Schleicher, Zhang was a physicist by training, a man of numbers with an eye for evidence. But the twenty-first century was no country for old scientists. 'Society is changing and our living conditions and status are improving,' he said. In his youth, he and his classmates had learned Russian. Now he advocated that the science of computing be included in the curriculum – although not coding, which he felt was simply 'a language between robot and computer'. The next generation needed to learn to deal with people. 'An American student or British student stands in front of the world stage and they show you great confidence,' he said. 'They don't care about academic performance.' He thought Shanghai's schools could learn something from that. But it was hard to define, difficult to measure. That was the point.

The next generation was no longer motivated by the pure struggle to better their lot in life economically. 'Every student is going to be the owner of his own life,' said Zhang. Teachers would need to think and act differently. New approaches would build on kids' interests and encourage their individuality. A one-size-fits-all system centred on the average was good, but it wasn't enough. Zhang's approach had brought about huge gains for millions of kids. Now it was time for new ways of thinking. It was that impulse that had taken me to visit Andreas Schleicher, and the same one that had led Schleicher to Shanghai. Zhang added, 'I cannot be the designer of tomorrow.'

Our New Religion

'In a changing world, remaining ahead is something that does not come by itself,' said Andreas Schleicher as we reached a dead end within the labyrinthine new HQ. Aptitude in data analysis didn't correlate to the simpler task of navigating corridors. A neatly coiffed head rose meerkat-like above a computer terminal to direct us to the lifts. 'Success is never for ever really. Complacency has high prices.' As we whooshed down to the lobby, I thought about this idea. Data was inescapably part of our future. It revealed new truths, shook lazy assumptions, showed us that we could be better. 'If you've done a lot of good things in education you're going to see that in the proof of the results,' he added. Out on the wet street, his wheelie-case clacking over uneven cobbles, I found myself seduced by his zeal. His data was fuelling a global surge in quality education. He wasn't just another person with an opinion.

'It is still striking to see how much progress is achieved in some parts of the world.' Many of today's top-ten countries hadn't been there a decade earlier. Shanghai and Vietnam had shown you could get *better* at teaching millions of kids, while others, 'like the US – that *was* number one in education – and is now barely scraping around the average', had proved you could be left behind. In Schleicher's hands, with his scientist's mind and reformer's faith, the data lit the path towards the future of learning. He pointed to efforts to bring back grammar schools to the UK, where 'immediately you have people saying, "well show me where it has actually worked, what is the impact on equity?"' It was *obviously* better to base things on proof. The only danger was that once you started measuring, people obsessed only over that which was measurable.

'If we don't feature those character qualities on our metrics,' he said, mentioning our elite private schools, 'it means that poor children will never get them.'

We had to frame data as a tool. Schleicher couldn't countenance decisions unsupported by evidence. But at the same time all measures were imperfect. No matter whether you agreed with them you were bound into the race. At Walworth, hapless and exhausted, I'd no longer asked whether the students of 10X4 were great readers or writers, but whether they were greens, yellows or reds. In Shanghai and Korea I'd seen a similar paradox. Kids were acing their tests, but the systems seemed more interested in their data than in them. The Suneung and Gaokao force-fitted the soaring faculties of kids' intelligence into a series of tick-boxes. We knew data was an image, 'a mirror' as Schleicher had put it, but it felt like sometimes our systems of learning didn't. They had learned to run on numbers.

As we boarded the Metro at Pont de Sèvres among the crush of Friday evening commuters, I asked Schleicher about this risk. Yuval Harari, the futurist author of *Sapiens* and *Homo Deus*, argued that the scions of Silicon Valley had introduced to our world a new religion of 'Dataism'. 'In its extreme form,' he wrote, 'proponents of the Dataist worldview perceive the entire universe as a flow of data, see organisms as little more than biochemical algorithms, and believe that humanity's cosmic vocation is to create an all-encompassing data processing system.'[18] Was that where our schools were heading? If the world wanted compliant, able white collar workers to drive basic economic productivity, the approach worked. If you wanted creative human beings, it was a misstep.

'We still have a very industrial work organisation,' agreed Schleicher. 'We expect the boss of the factory to know what every worker is doing.' To adopt a managerialist approach in our schools was to waste human potential. Instead, we ought to aspire to build systems that served to make *us* smarter and happier. *We* created knowledge and insight. People weren't supposed to serve systems. They were meant to serve us. In schools this was even more the case. 'You know, actually you don't find any sector of our industries that has as many highly qualified people working for them as education. Every teacher has a university degree. Find me any company that can afford that.' Huge potential existed in the minds of teachers and learners. 'But we're not using these people for knowledge creation; we're using them to implement prefabricated wisdom.'

As he gripped the handrail for balance I noticed the dormant Apple Watch around Schleicher's wrist.

'It's actually very smart,' he said, raising his voice as the train powered along. 'It's got all my emails, appointments on there. It has Google Maps.' 'I got it as a present, and then suddenly after three days I heard it say to me, "Leave now to make your appointment with Mark Tucker." And so I asked myself, so it knows my appointment because my secretary put that in my calendar! How does it know where I actually am? Of course, it's all in the GPS. The amount of data that they collect is incredible.'

I reflected that our devices now held more than the blunt statistics of Adolphe Quetelet or the age-old multiple-choice answers of the Suneung. Data was up-to-the-second accurate and entirely personalised. Schools were catching on. At Rocketship they gathered and interpreted data on every keystroke of the purple-shirted Rocketeers. Certain MOOCS were minutely tracking user progress through a series of world-class online courses. Free to access, they could earn revenue by offering detailed employee profiles outlining the skills potential hires had mastered, but also how long they took to acquire them, where they failed and repeated.

It mattered *who* used the data and what they used it *for*. In the hands of able humans it could fuel judgement. In the hands of an algorithm – or an authoritarian state – data became a tool of discipline and punishment. The Chinese government was already well on the road to digital authoritarianism. Building on the tradition of the *dang'an* folder* it held on each citizen, containing medical,

---

* The *dang'an* is a record held by the Chinese government on the 'performance and attitudes' of every individual citizen in the country. It contains a photo, along with information on a person's physical characteristics and employment record. It also contains academic reports from primary school to university, society memberships, criminal record and political history.

educational, employment, travel and criminal records, the state had begun testing in 2010 a big-brother-meets-big-data single database social-credit system that would track everything from salary details to whether you'd visited your parents enough recently.[19] The government had the data already. Only software issues were holding back the government's *Minority Report* desire to use the data to predict and pre-empt antisocial behaviours.

It was unsettling. Today you had a second chance if you flunked your exams. But if every interaction were recorded, your data really might become your destiny. The kids of Korea had been stressed enough about a single exam. What if they were on the hook for every test they'd ever taken, every piece of schoolwork they'd ever done? If every activity, every interaction, every success, every failure, was logged in the block chain of your life, an inescapable, immovable record, wouldn't every day at school – and every moment beyond it – risk taking on the brutal normalising power of the Suneung? That was the irony of the examined life in the early years of big data. The roots of exam preparation ran so deep that even a move away from high-stakes testing might result in even greater pressures.

'We are already becoming tiny chips inside a system that nobody really understands,' wrote Yuval Harari. Was that where schools were heading? Ministers in the UK were pushing high-stakes testing ever earlier into kids' live, jumping on the managerialist bandwagon a hundred years late. In the US, academics had shown the perverse effects of testing on the content and form of all kids' learning.[20] I thought of Seung-Bin's exhausting progress to his reckoning with the Suneung and how light he'd seemed a month later, reading for pleasure for the first time in a decade. We had to trust our teachers to resist the brutal logic of the system. Our kids too. We needed exams to evaluate our progress, needed data to light our path. But somewhere, somehow, we needed to stop teaching to the test.

'I would worry if it becomes high stakes,' said Schleicher of his PISA exams. They were intended as a tool, a global public good, not to reinforce hierarchies, but to undermine them. But he sensed that people were turning inwards. Schools were our only bulwark against a radically changed world. What really mattered – what set you apart as a human – was whether you actually understood the social, political and cultural phenomena in the world around you 'and learn to capitalise on them'. A truly caring system didn't see data as an end in itself, but as a tool to advance and realise our innate learning potential. This was the idea we needed to fix in our minds. Though it felt a long way off to me, Schleicher was hopeful. Data was our servant, and he served it.

The train pulled in at Trocadéro station and he turned to salute me farewell before loping off against the tide of Parisian commuters, a lone data ranger fighting the global education orthodoxy.

As I swept along on the Metro towards the centre of Paris and my Eurostar back home, I reflected on how we might use his findings to make our schools more caring, more human. Evidence was vitally important, but we had to put it to the right use. Our natural born learning potential was multidimensional, couldn't be force-fitted into a data-shaped mould. We had to learn to take equally seriously the things that couldn't be measured, see what it might look like to put well-being first, or to value kids thinking together about the world around them and figuring out what they might do to change it. It was with those thoughts in mind that I went on to New York, the next stop in our journey. Twenty years earlier, a US head teacher in the Bronx had seen his pupils perform brilliantly in their state assessment, then fail to carry their successes into life beyond school. How, he'd asked, could he set them up to thrive in the real world?

CHAPTER 8

True Grit

Character-Building Stuff

> You ought to thank me, before I die, for the
> gravel in your guts and the spit in your eye,
> 'cause I'm the son of a bitch that named you Sue.
>
> Johnny Cash

Knowledge is Power

Dave Levin had a problem. In 1995 he'd opened a new middle school in the South Bronx with a promise. Come to KIPP Academy NYC at ten, graduate college at 21. It was an outrageous claim. The data showed that at most one in ten kids from the neighbourhood were likely to earn a degree, while nationwide just seven in 20 Americans of any background made it through higher education. And this was the South Bronx, one-time crime capital of New York City, birthplace of hip-hop and b-boys, Afrika Bambaataa and Grandmaster Flash, not an upstate private school on the leafy banks of the Hudson. College wasn't for these kids. But Levin had believed. And trailing door to door with his goofball enthusiasm and oversize shirt and tie – hey, it was the Nineties – he'd persuaded enough parents to join him. Three in four students would make it to graduation day went the promise, just as they did upstate.

Dave had co-founded KIPP a year earlier with fellow teacher Mike Feinberg. Fresh-faced Ivy-Leaguers out of Yale and U. Penn, the pair had met in Houston, where both taught elementary school as corps members of Teach for America, a programme launched five years earlier to encourage the brightest young minds in the US to take on the country's crisis in educational inequity. After some early struggles, the eyes of the two teachers had been opened to the practices of Harriet Ball, a six-foot-one veteran black elementary school educator with her own version of the 'Three Rs', 'repetition, rhythm and rap'. Her radical approach, comprising songs, chants, games and raucous jokes, became their blueprint for motivating and educating inner-city kids.[1]

'Harriet Ball was a rock star teacher,' recalled Levin, 'she came into my room one day and in forty-five minutes taught what I had failed to teach in three months.'[2]

They began to apply her practices in their classroom. The kids were entranced. They loved it and their progress was rapid. The two young educators were sold and decided to systematise the approach.

One of Ms Ball's best-loved chants was 'You gotta read, baby, read. You gotta read, baby, read. The more you read, the more you know, 'cause knowledge is power, power is money, and *I want it*.' It gave them their name: the Knowledge is Power Program. Feinberg stayed at KIPP Academy Houston. Levin returned to New York.

The first two KIPP schools became laboratories of high achievement for low-income kids. On average their fifth-graders came in two years behind, reading, writing and doing maths at a third-grade level. They also had complex behavioural needs – these were vulnerable kids from tough urban backgrounds. Everyone sat back and waited for the two educators to fail. They were crazy. But Levin and Feinberg had zeal. And they clung to a powerful mantra. *Whatever it takes*. Kids would get to school at 7.25 a.m. each day and stay until 5 p.m. They'd attend class every Saturday morning and give up a month of their long vacation, which Levin called a relic of an agricultural age when kids took part in the harvest, to attend summer school. The equation was simple. More time in school meant more learning. And KIPPsters – the schools had their own lexicon – spent about 70 per cent more time in class than their peers, just like they did at KSA in London, which was inspired by their work. Aiming for the very top – to overtake, not just catch up on their fifth-grade deficit – meant they needed all the time they could get. At the entrance to each school was a slogan: *Climbing the mountain to college.*

This is what it would take: KIPP turbo-charged learning with Ms Ball's model. Kids chanted. They recited facts and knowledge over a break-beat. They drilled routines. The schools were run according to a simple philosophy. *Work hard, be nice.* It was written on the walls, printed on the back of kids' shirts, just like it was on Ifrah's. And they did work hard. Teachers worked 60-, 70-, 80-hour weeks. On the first morning of the school year, they gave out their mobile-phone numbers to students. If they needed help with homework, they called up a teacher, no matter the time of day or night. Underpinning this was the 'Contract'. Signed by students, parents and educators, it was a solemn promise by each to do whatever it takes. 'The kids come to school with a hundred and one reasons why they're set up for failure,' Feinberg later explained, 'that means we need at least a hundred and one solutions for how to set them up for success.'[3] There would be *no excuses.* This meant military levels of discipline. Kids were taught how to sit, how to hand out books, how to walk corridors. The expectations of their behaviour were as high as for their grades. Each class was named for the year it would graduate college. Those first cohorts were the Class of 2003.

The hard work paid off. In 2000 the US current affairs show *60 Minutes* profiled KIPP as its first eighth-graders graduated.[4] The results were sensational. In just four years, fifth-graders who had been two years behind at the start of middle school were now mastering early high school algebra classes. Every student from both academies was going on to a prestigious private or public high school, a good sign that they were on track for college. The 69 kids at KIPP Academy Houston amassed more than $1,000,000 in scholarships between them, bringing the school national fame. Against the back-drop of a crisis in public schools, here were two outlier institutions where poor kids were outperforming their richer peers. The Fisher family, founders of GAP, called up and asked them how many millions of dollars it would take to scale KIPP nationwide.[5]

They opened new schools and grew into a network spanning the US. KIPPsters continued to chant and to SLANT (Sit up tall, Listen, Ask questions, Nod your head, Track the teacher). And they continued to excel, making progress each year far beyond what was expected. By 2006, Mike and Dave were sitting in armchairs on the stage of *Oprah*, energetically telling the story of KIPP and bringing tears to their host's eyes as she watched a video of a young black boy from the South Bronx proudly proclaim 'I want to be smart'.

Through all of this, they stayed in touch with those first classes, bound together by the Contract. They picked up the phone to their former students, guided them through high school, helped them navigate family issues, kept them climbing that mountain.

Most of them made it. Nearly every member of the Class of 2003 graduated high school and won a place in college. But then something had happened. Some kids had gone on to top high schools and then to four-year colleges, far exceeding the expectations society held of kids like them. But others hadn't fared so well. For one reason or another they'd dropped out along the way. Money troubles. A family crisis. The need to get a job. Failing to get on with a teacher. Not liking college life. Falling behind. Feeling out of place. Though nine in ten of that first class graduated high school, just one in five – eight students in total – made it to the end of a four-year degree. It was still three times the rate of other kids from their backgrounds, but a long way short of the 75 per cent dream.

For their critics, it was proof that they'd aimed too high. But for Dave Levin it was a problem to be solved. The kids had made it *to* college, now they needed to get *through*.

In the mid-Noughties, he started to study the problem, looking at the stories and circumstances of those who had made it and those who hadn't. A minimum level of academic attainment was a non-negotiable. Reading, writing, maths and science had to be in place. But this wasn't the clincher. Kids with great academic records had dropped out too. Something more intangible seemed to matter and he landed on a hypothesis. 'When we looked at the kids who were graduating we noticed that not only did they have these academic skills, they also had a set of character skills,' he later recalled in a radio interview. 'They were more likely to persevere, they knew how to reach out and build relationships with professors and they knew how to ask for help before they got in such a deep hole.'[6]

Levin turned to a burly, balding professor at the University of Pennsylvania. Dr Marty Seligman was a former president of the

American Psychological Association, founding father of Positive Psychology and a perfect match for solving Levin's problem. He often talked to educators and liked to begin with a pop quiz. First, he asked his audience to describe in two words or fewer what they most wanted for their children in life. *Joy! Purpose! Happiness! Love!* they would call out. Then, he asked, again in two words or fewer, what did schools teach? They always laughed, and then they'd tell him. *Compliance! Subjects! Process! Facts! Numeracy! Literacy!* He then paused for effect. Notice, he'd say, that there's no overlap between the two lists.[7]

Seligman, who'd spent his life helping the mentally ill get better (he'd shown experimentally, for example, that for many sufferers depression was a 'severe low-mood' that could be treated with therapy, rather than medication),[8] was now on a new mission. He aimed to use his techniques of 'Positive Psychology' to improve the well-being of all. Levin had read his book *Learned Optimism*, and called him up to arrange a meeting. It was little more than a hunch. If KIPP was going to make good on its promise, Levin needed to know what the character skills were that carried kids through college. This chapter tells the story of Levin's effort to solve that problem and how it came to change KIPP. It shows how taking care means helping kids flourish far beyond any ability to do well on a test, that as natural born learners our potential extends beyond even thinking and doing, to feeling. And it starts with a simple question: why did some succeed, while others failed?

The Content of Our Characters

Seligman's set-up is all too familiar to teachers. We know what matters, but are powerless to resist the terminal logic of a system dialled towards high-stakes exams. Today character education is increasingly important. Our era demands more of our human talents – empathy, creativity, sociability – and raises the spectre of a future in which we'll need others, like drive, determination and resilience. We're living under the shadow of an epidemic of mental health problems. In 2016, one in seven 16- to 24-year-olds in the UK experienced mental illness (anxiety, depression, panic disorder, phobia or obsessive compulsive disorder),[9] up to one in four women of that age. The World Health Organization estimated that 450 million people are currently affected by mental disorders globally and that one in four will suffer at some point in their lives.[10] Strength of character is now needed to stay sane, let alone thrive. The 'flow of novelty generated by today's market-based consumer societies is so strong', wrote Avner Offer, Emeritus Professor of Economics at All Souls College, Oxford, 'that higher levels of commitment and self-discipline are needed to ensure that long-term well-being is not sacrificed for short-term gratification'.[11]

It's not a novel concern. Character development always was at the heart of human learning. Two and a half thousand years ago, the students of Plato's Academy were more interested in how men should live than with the intellectual tools of rhetoric or logic. The field was known as ethics, or moral education. Aristotle, the Academy's most famous graduate, wrote one of his best-known works on the subject, the *Nicomachean Ethics*, a proto-parenting manual to help fathers raise sons. It was named for Aristotle's dad and firstborn and worked as a how-to guide for developing a virtuous moral character, which for Aristotle meant a balanced alignment of your passions with your reasoning mind. Most importantly, he intended his ethics to be *practical*, informing human behaviour.

At Walworth Academy, this aspect of learning was much discussed, but little understood. Pupils' character developed not because of a concerted scheme, but through the tireless efforts of individual teachers. Many of the staff wanted above all to raise good human beings, but while the techniques for securing GCSE grades were precision-engineered, our shared efforts at helping kids develop socially and emotionally were haphazard, consisting of the usual arms round shoulders, whole group lectures on behaviour and token Personalised Learning and Thinking Skills posters on display boards. We had no practical ethics, no twenty-first-century playbook. We didn't even know what to call it. As a pupil, I'd at least followed an unwritten code of sorts. Anglican in spirit and Corinthian in ethos, it had kept us running on the rugby pitch and sent boys haring for the army as soon as they left.

So how did you develop good character? Levin and Seligman

weren't interested in philosophising. They sought *impact*. KIPP cared about raising successful human beings, ones with more to show from school than test scores. But working on your stiff upper lip or debating the good life seemed passé. Today we were less concerned with what was right and more concerned with feeling well or figuring out how to perform to our maximum. Life coaches would soon outnumber philosophers or politicians. Perhaps they already did. Character in that context might be better understood as combining both well-*being* and well-*doing*. Happily for Levin, the growing consensus among psychologists in the US that you could broadly (albeit artificially) boil down all personality traits to some variation on a 'Big Five' of agreeableness, extroversion, neuroticism, openness to experience, and conscientiousness gave him a starting point. Moreover the experts had begun to develop some practical tools to help take better care of kids.

One of Seligman's protégés was making waves studying the last of these five. Seligman put Levin in touch.

You're Getting Off First, or I'm Gonna Die

Selection for West Point, the US Army's Military Academy in upstate New York, is – unsurprisingly – tough. Of 14,000 hopefuls who apply each year, just in one in 12 is admitted. Among them are the finest minds and bodies of their generation, varsity athletes and 4.0 GPAs.* Yet for all the rigours they suffer to get there, one in five drop out before the bugle sounds for graduation day. More remarkably, most of those make their melancholy march down washout lane before their first summer is even through. Angela Duckworth, the flinty University of Pennsylvania psychologist employed to investigate the phenomenon, sums it up neatly: 'Who spends two years trying to get into a place and then drops out in the first two months?' The answer lies in the nature of those eight weeks. The brochure

---

* US students receive a 'grade point average', or GPA, at the end of high school. You are awarded a score from 0 to 4 in each class, and then scores for all classes are added up and an average score found. 4.0 is the highest GPA, equivalent to achieving straight As.

describes them as 'the most physically and emotionally demanding part of your four years at West Point'. They call it 'Beast Barracks'.[12]

As Levin was pondering his problem, Duckworth was two years into a psychology PhD investigating adolescents and achievement. Now a TED Talk superstar and MacArthur Genius who was awarded a $1,000,000 fellowship by the foundation for her ground-breaking research, she had until then led a peripatetic career. After growing up in Philadelphia to Chinese immigrant parents, she studied neurobiology at Harvard and post-grad neuroscience at Oxford. As for so many of the very ablest minds, particularly those without a clear game plan, her next stop was McKinsey, the blue chip consulting firm. Uninspired, as with so many of the very ablest minds, by the corporate world, at 27 she swapped her suit, heels and PowerPoint decks for a pair of sensible shoes and headed out to the Lower East Side to become a middle school maths teacher.

'It was consulting,' she says, 'not teaching, that was the detour.'[13] Human development was where she found her purpose. In her classroom in Alphabet City – dominated then by housing projects rather than hipsters – she reached the same conclusion as Dave Levin. It wasn't the kids with the most natural ability who scored the best grades in her classes, but the ones who worked the hardest, just as Anders Ericsson had argued, and Ifrah has shown us. 'Aptitude did *not* guarantee achievement. Talent for math was different from excelling in math class.' Her realisation – *I'd been distracted by talent* – prompted a further career change. If ability didn't predict high achievement, then what did? Leaving the classroom at 32, she set off to find out.

Her research led her to West Point, where military psychologists had been studying candidate attrition for years. Their best proxy for predicting success was the Whole Candidate Score, a weighted average comprising exam results, an expert evaluation of leadership potential and measures of fitness. It was a shorthand for a candidate's talent, but it had no predictive power in determining who might drop out. Interviewing an army scientist one day, Duckworth had an epiphany. 'Recruits were being asked, on an hourly basis, to do things they couldn't yet do.' Talent couldn't help them. They had to draw on some other source. Faced with possible failure, some buckled.

But the others didn't. They had a 'never say die' attitude that carried them through. This was true of other high achievers she'd studied – superstar journalists, academics, athletes and business people. 'These exemplars were unusually resilient and hardworking. Second, they knew in a very, very deep way what it was they wanted. They had determination, they had *direction*.' Put simply, 'They had grit.'[14]

Duckworth hastily assembled a first Grit Scale. Subjects would rank themselves from 'Very Much Like Me' to 'Not At All Like Me' on a series of statements like 'I have overcome setbacks to conquer an important challenge' or 'I am diligent. I never give up.' She administered the Scale that July to 1,200 recruits on day two of Beast Barracks. After first checking for correlation between Grit Scale Score and Whole Candidate Score (there was none – talent and grit were unrelated), she sat back and waited for the summer to pass. By the end of it, 71 candidates had dropped out.[15]

Though stayers and leavers had indistinguishable Whole Candidate Scores, 'grit turned out to be an astoundingly reliable predictor of who made it through and who did not'. Duckworth repeated the Scale in 2005, with 60 cadets dropping out overall. Again, her findings held. She repeated the study on salespeople, Green Berets, Chicago schoolkids, even spelling bee contestants. Every time, her findings were confirmed. Grit, not talent, predicted who would succeed and who would fall away.

At a recent talk, I saw her illustrate this insight with a YouTube clip of actor Will Smith on the Tavis Smiley chat show. 'I will not be outworked,' he says, fixing his gaze on the host. 'You might have more talent than me. You might be smarter than me. You might be sexier than me. You might be better at nine out of those ten things. But if we're getting on a treadmill together, you're getting off first, *or I'm. Gonna. Die.*' I could see how an attitude like that might strengthen a kid's well-*doing*, but I was a little concerned about its effect on their well-*being*. It was with this thought in mind that I'd decided to journey to the East Coast of the US, fast becoming the global nerve centre for a character education revolution.

One KIPP school in Harlem, New York, was taking Duckworth's insights particularly seriously. How did you keep kids on that treadmill?

To Infinity and Beyond!

On a wet New York morning on West 133rd Street caterers were arranging breakfast platters under an awning for the cast of *Law & Order*. They often filmed the cop show in Harlem and you could see why. Iron bridges loomed over a barely marked road. A line of trucks tunnelled a biting Hudson wind along the sidewalk. Tough-looking men in yellow hardhats carried loads onto a construction site. 'It's the kind of place where bodies wash ashore,' joked a teacher later. But it was a mirage. Those days had passed. Kids bustling past the trestle tables in their down jackets and Peppa Pig backpacks looked to a new dawn. The building going up was a high-tech Columbia University facility. Across Twelfth Avenue the Fairway supermarket and Harlem Bierstrasse Beer Garden signalled regeneration. And the school they were entering alluded to a future of unlimited possibility.

KIPP Infinity Harlem is one of the now 200 KIPP schools serving 80,000 kindergarten to high school students in 31 US cities. True to their founders' mission, all of the schools serve low-income communities, 96 per cent of admissions are African-American or Hispanic, 88 per cent of kids are eligible for free or reduced lunches, and 10 per cent qualify for special educational services. Though the data still says these kids shouldn't succeed, they do: 94 per cent of KIPPsters graduate high school, 81 per cent start college and 44 per cent graduate a four-year programme, more than double the rate of the first cohort. Though they're still some way off Levin's ultimate goal of matching the rate at which kids from the top income quartile make it, the numbers are ahead of national averages for all kids of any background, which stands at 91 per cent, 64 per cent and 24 per cent respectively.

I pushed my way inside out of the freezing rain.

In a large wood-panelled auditorium on the third floor, 350 fifth-, sixth-, seventh- and eighth-graders stood for their Monday morning pledge. They were dressed in maroon, racing green, navy and black polo shirts, and great sneakers. 'We believe,' they said in unison, 'that with desire, discipline and dedication, we can change our world and our place within it. There are no shortcuts. There are no limits. As a team and a family, we will either find a way or make one.'

Twenty-two years on, here were more young teens intoning that dream. Climbing the mountain to college. Doing whatever it takes.

'We *are* KIPP Infinity.'

It felt a little like a church.

At the front of the room, Principal Allison Holley took the stage. That weekend had seen the inauguration of President Trump. Among the students, of whom one in four were African-American and three in four Hispanic, it was a cause for concern. She shared photos of KIPP Infinity graduates taking part in the New York Women's March, which had ended outside Trump Tower. 'You remember them?' she asked. 'We saw one part of democracy on Friday, and another part on Saturday.' Principal Holley wanted them to know that although it was a tough world out there, they had power, ought never to give up. She followed it up with a picture of the boys' bathroom from the previous Friday, strewn with paper and litter. 'I am going to give you a week to deal with it,' she said. They had responsibilities too.

Hurrying down the hallway in my wet shoes, I'd noticed how central these ideas were to KIPP Infinity's ethos. 'We are what we repeatedly do,' suggested Aristotle from the walls. 'Excellence, therefore, is not an act, but a habit.' Above the sticker-laden student lockers – most prized were Millionaire badges, awarded to kids who'd read that many words – were banners plotting the steps to good character. 'One. We don't want to get into trouble. Two. We want a reward. Three. We want to impress someone. Four. We follow the rules. Five. We truly care about the rights and feelings of others. Six. It's our code. It's who we are.' They thought deeply about this stuff. But anyone could put up posters. I had done it myself. How were they really preparing kids for the world?

I headed to Mr Griffith's classroom to find out. Twelve years earlier he'd been the school's founding maths teacher. It was around then Levin had contacted Seligman. At the meeting, the psychologist had derailed the conversation, introducing Levin to a new book he'd written with Christopher Peterson. *Character Strengths and Virtues: A Handbook and Classification* was a doorstop. At 800 pages in length, it was intended as the first 'science of character', analysing virtue throughout human history, from Plato to Pokémon. It proposed six strengths common to all races, classes, genders, cultures, eras or places – 'basically all the things that divide us', said Levin.[16] This was not, according to Seligman, a set of 'ho-hum findings that every Sunday school teacher or grandparent already knew', but a veritable 'manual of the sanities',[17] a companion volume to the APA's infamous encyclopedia of mental illness.* Or *Morality for Dummies, 2.0*.

Under the six strengths of 'Wisdom and Knowledge', 'Courage', 'Humanity', 'Justice', 'Temperance' and 'Transcendence' were listed 24 positive characteristics. 'We think of them as a combination of strengths and skills because traits imply you're tall or you're short,' says Levin, 'you can't change them. But self-control and curiosity and social intelligence, those aren't fixed.' They had chosen seven that KIPP would prioritise. I saw them on the wall of every classroom: *Self-Control, Grit, Optimism, Social Intelligence, Gratitude, Curiosity* and *Zest*. Levin dreamed of a world in which college admissions tutors would take your CPA – character point average – just as seriously as your GPA. But it had been hard to imagine a reliable system of measurement. They settled on the idea of a Character Growth Card. Kids would be regularly graded from one to four on each of the seven strengths. Mr Griffith had been there for the launch.

In his classroom, a sign above the whiteboard read, 'Progress: at what cost?' in six-inch-high letters.

In their black school-issue hoodies emblazoned with *Infinite Character*, the eighth-graders were reading up on the muckrakers, fearless journalists like Upton Sinclair and Ida Tarbell who'd advanced American democracy in the 1920s by upholding the Fourth Estate against shady cabals of corrupt industrialists. They were also learning to annotate, painstakingly highlighting appropriate passages. Hundreds of collectible action figures, X-Men, Batman,

---

* The APA's *Diagnostic and Statistical Manual of Mental Disorders* is a standardised list of 297 mental disorders that has been praised for bringing shared language to the treatment of mental illness and criticised for contributing to a huge surge in the medication of things like ADHD and depression. Eleven per cent of four- to 17-year-olds in the US are diagnosed with ADHD, with 6 per cent on medication, according to America's Centers for Disease and Control, compared to 3 per cent and 1 per cent of UK kids respectively. (Sarah Boseley, 'Generation meds: the US children who grow up on prescription drugs', *Guardian*, 21 November 2015.)

Superman, each avatar pregnant with its particular character strengths, kept a vigil from their glass case to our rear. A two-foot R2D2 perched on a stool. Class finished with a clip from *Law & Order*. It was a modern slaughterhouse scandal, after Sinclair's *The Jungle*. The content was political, character-building.

The kids filed out and I caught up with Mr Griffith. What was KIPP doing to develop character?

'Do I give a lot of optimism stars? No.' Mr Griffith told me about KIPP's 'pay-check' system. Students began the year with virtual dollars credited to an account. Stars meant extra dollars, while demerits meant a penalty. 'Pay-checks' were handed out each Wednesday, 'like death and taxes'. The aim was to make it to the end of the year in greater credit than you started. Short term you'd be rewarded with access to trips, long term you could work your way off the pay-checks and become *salaried*. Once you achieved that status you were off the pay-check system for good, barring any serious lapse, your good character on secure foundations.

KIPP had been notorious for the high demands it made of pupil discipline. Many kids didn't make those trips.

Initially, Character Report Cards were given to kids four times a year, with teachers sitting down to reflect on a kid's strengths and give them a grade. But recently they had been phased out, as teachers felt they didn't accurately reflect a student's character. Mr Griffith felt they had been more arbitrary ('You can imagine being asked to do ninety of those on a Tuesday'), whereas pay-checks felt really valuable.

'In advisory, we will have all students look at their pay-check. What are you struggling at? What are you doing well at? They'll sit with their KIPP Circle and try to figure out goals to improve their pay-check average.' The goals would always be explicitly tied to character strengths. A student aiming to improve their pay-check might 'bench' themselves in recess (bench was KIPP's word for detention) in order to do some extra reading. This was an exhibition of self-control. When teachers rewarded them, the idea was 'not just [to] give them a star, but also to articulate what they're doing'. It seemed like KIPP was hoping that students would be able to develop a more dispassionate view of their behaviours, to identify and then think about them. It was a form of cognitive behavioural therapy.

I could see why the focus tended towards grit and self-control. These were the more visible ingredients of academic success and seemed somehow better aligned with the KIPP mission. As the school basketball coach, Mr Griffith saw how hard it was for a kid to succeed even when their motivation was high. 'Even with this thing that you love to do, are you showing self-control to practise the right way? If you're bad at free throws why are you shooting threes?' Climbing a mountain took grit above all. Though the school emphasised creativity, optimism and joy, grit was what they were known for. Though now close to half of KIPP's graduates *were* making it through college, I could understand why some critics had labelled KIPP as the darlings of corporate America. The students would make great workers. They were even used to pay-checks. I couldn't help thinking of Seung-Bin back in South Korea striving through 14-hour school days.

'I don't think we found a way to really see when a student shows optimism or gratitude,' said Mr Griffith. That felt a way off. Less instrumental in their careers perhaps, these felt fundamental to kids' well-being. Was KIPP too focused on self-control?

To Scoff or Not to Scoff

In 1968 the Viennese-born psychologist Walter Mischel set out to understand self-control. In his Palo Alto kitchen, he'd observed his young daughters quickly passing through the impulsivity of their toddler years to become more rational, patient beings. What was happening in their heads now that they could delay gratification? He devised a simple experiment. In it a researcher would place a marshmallow (or some other treat) in front of a preschooler and give some simple instructions. When they left the room, the child could either eat the marshmallow (if they did, they should ring a bell), or if they could hold off eating it until the researcher returned (they didn't specify how long they'd be), the child could have a second marshmallow. Having used his daughters as guinea pigs, Mischel took the experiment to Bing Preschool, a purpose-built unit at Stanford University. His daughters' playmates would be his test pilots.[18]

The Marshmallow Test is now world-famous, with YouTube hosting endless clips of kids contorting themselves to not take the treat, due to its incredible predictive power. Mischel studied the Palo Alto preschoolers for over five decades with astounding results. How long a child waited to eat a marshmallow as a kindergartener directly correlated to their success in life. Two-marshmallow maestros did better in school, earned more in their careers and were healthier and even happier than their more impulsive peers. The snack-scoffers were much more likely to experience negative outcomes like ill health or jail time. Over the years, Mischel repeated his study with other kids in a variety of contexts. The results remained constant. Willpower was destiny.

Mischel didn't stop there. During the tests, he'd noticed that those who managed not to eat the marshmallow used inventive strategies. The psychologist noted that they 'cover their eyes with their hands or turn around so that they can't see the tray, others start kicking the desk, or tug on their pigtails, or stroke the marshmallow as if it were a tiny stuffed animal'.[19] Perhaps other kids could learn to do the same? He began showing his subjects how to cheat at the game. The most effective strategy was for them to put the marshmallow in an imaginary picture frame and treat it as a virtual object. Kids who had previously waited only a minute before caving in could now stretch their patience to as long as 15. You could *grow* self-control.

The marshmallow sparked what Mischel calls 'hot emotions' in the kids, foreshadowing the work of B. J. Fogg. Like Homer Simpson pining for a Duff, they all *desired* the treat – the feelings were automatic, impulsive and deeply rooted in their neural pathways. Resisting the temptation depended on their ability to 'cool' that hot impulse, deploying the prefrontal system to override the urge. Maestros managed to pay attention to the treat in a different way, considering its shape, texture or colour for example. Even imagining it was something else. The repercussions of this skill were wide-ranging. 'If you can deal with hot emotions,' Mischel told the *New Yorker,* 'then you can study for the S.A.T. instead of watching television. And you can save more money for retirement. It's not just about marshmallows.'[20]

Excited by the possibilities of these findings, Levin was keen for

Duckworth and Mischel to use his KIPPsters as the first test subjects outside the lab setting. Might teaching fourth- to eighth-graders at KIPP how to resist the marshmallow result in a transferrable growth in self-control? Mischel thought it unlikely. He knew that character was really made up of habits, and that habits took years to shape. But he was willing to try. It was notoriously difficult to secure permission to work *in the wild* and he felt the experiments could one day lead to a revolution in our approach to school. 'What we're really measuring with the marshmallows isn't willpower or self-control,' he continued. 'It's much more important than that. This task forces kids to find a way to make the situation work for them.' He wanted to understand how we could equip kids with those strategies.

'They want the second marshmallow, but how can they get it? We can't control the world, but we can control how we think about it.'

In Peace, Out Worries

In a low-lit room panelled in soft foam, 25 first- and second-graders assembled themselves cross-legged on pastel-coloured cushions. Beside me, Naomi closed her eyes and laid her hands on her knees. Alberto grinned up at the ceiling as he fidgeted through a few poses, and settled on the starfish. The space, with its lilac, pistachio and powder-blue walls, had a calming effect. I welcomed it. The day had begun at 5.30 a.m. in West Hartford's New York Sports Club, where Principal Julie Goldstein ran the morning spin class. 'The hardest part is over,' she'd yelled over the Skrillex, 'you turned up!' We'd carried on pedalling ('If it feels like you're slowing down, then you *are*!') for 45 minutes. A character class for grown-ups, I'd thought as I sweated up an imaginary hill. I *could* do it. But for all Ms Goldstein's joie de vivre I kept thinking about Angela Duckworth. The darkened room had resembled a club, but instead of group abandon it bred self-discipline. Was every school signing up to a future defined by the Jane Fonda doctrine?

Back at the edge of the rug, Ms Soto-Gomez – presidential-debate ready in a Hillary-esque blue trouser suit – took a long deep breath. She was the school's Mindfulness director, a psychiatric social worker of thirty years' experience. 'Peace starts with yourself and grows,' read a note outside her office. 'Always believe that something wonderful is about to happen,' offered another. Where the Dalai Lama simply *said* that if we teach every kid to meditate, we could eliminate violence from the world in one generation, Maritza Soto-Gomez was *doing* it. Her trademark Mindfulness Program ran from pre-K through to eighth grade, following the teaching of Jon Kabat-Zinn, a physician and founder of the Stress Reduction Clinic. 'You can't stop the waves, but you can learn to surf,' he'd written. At this young age, the kids were just paddling in the shallows. In today's class they'd be learning to recognise feelings.

'Hands on your knees,' she began in her soft Puerto Rican tones. 'Assume your mindful position.' The circle of pint-size yogis straightened their backs. Ms Soto-Gomez handed over to Mariela. An eighth-grader who'd been meditating at school for over six years, she was now a mindfulness coach, something the school encouraged. Their students would be disciples of meditation, taking these techniques forth into high school and the world beyond.

'We're just focusing on our breathing,' she continued.

The kids inhaled in unison, some a little ostentatiously, some with their eyes still secretly open.

'In, peace,' she intoned.

'Out, worries,' they breathed.

Two years earlier, Breakthrough Magnet in Hartford, Connecticut, was voted number one of 3,400 such schools in the US. Magnets had emerged to combat the segregation rife in public education in the 1960s. The aim had been to create institutions with a specialist area of focus – science or performing arts, say – and then situate them in the suburbs where they would attract affluent neighbourhood kids who would be joined by students bussed out from the inner cities. To secure funding, Breakthrough – most of whose kids were from low-income African-American and Latino families – ensured that at least of one in five of its students were white or Asian-American. The school specialised in character education and was named for it. Throughout the corridors kids were reminded of the acronym BRICK. Kids would learn to change breakdowns to 'breakthroughs'; take 'responsibility' for their well-being; learn 'integrity'; find opportunities to 'contribute' and increase their 'knowledge'. Mindfulness was the key.

Earlier, Ms Soto-Gomez explained that 'the same amount of effort we put into the reading and math, we put into knowing who they are, developing social skills, to understanding their feelings in the moment. To understand what do I need to do, without judging myself or judging others. To know that it's OK to ask for help.' I'd seen it in every classroom. The walls were festooned with helpful hints prompting pupils to be BRICKs. Each had a mindfulness corner with chair and desk, comfy cushions and mandalas to colour in. In the music room this even included a fully functioning minia-ture Buddhist fountain with faux rocks and foliage. 'Most of us eat, breathe and live this,' added Windy Peterson, the Special Education lead. 'It's not just teaching, it's who we are.' It felt New Age, but there was much more to it than greeting card platitudes. The prac-tices were rigorously applied, scientifically proven. Teachers used mindfulness as a tool to develop their kids' metacognitive abilities, as a few pioneers were also doing worldwide. It was a game-changer.

Shifting her weight on the pastel cushion, Ms Soto-Gomez picked up from Mariela. 'When we're sitting in this position we might feel a little stiff. So, arms up. Stretch your arms. Swish your hands.' The kids pantomimed, stifling giggles as they reached. 'Now, big breath. Breathing in. Out. In. Out. Now slowly, slowly, hands going down. Feel your whole body. Hands down to your knees.' Breakthrough defined mindfulness as the 'moment-to-moment, non-judgmental awareness of what's happening now: your breath, your body, your thoughts, your emotions, and your environment'. The practices were rooted in positive psychology, particularly the ABC model developed by founding father of cognitive behavioural therapy, Albert Ellis. The model had been proven to reduce anxiety, depression, adjustment disorders and conduct problems in kids. As part of Marty Seligman's Penn Resilience Program, it had reduced symptoms of depression in young people for up to twenty-four months after completion of a course.[21] At Breakthrough the rituals softened the science for the inner-city elementary school kids – as yoga-teachers did for the gym-goers of West Hartford – but the basic structure remained. You built awareness of activating events, thought through the behaviours they prompted in you and analysed the consequences. The goal was to become practised in self-awareness, so you could distinguish the steps, then tame emotional reactions to events.

In today's class, Ms Soto-Gomez's mini-disciples would learn to recognise feelings. To the sound of panpipes, she checked they understood the meaning of feelings, then introduced a game. The yogis remained hushed.

'I am going to say a word, then ask how you feel when you hear that word. The word that I am going to choose is *recess*.' A number of hands went up. 'Happy,' offered Maya. 'Joyful,' said Arianna. But what about when recess is over and I need to come back to class? Ms Mullings, the class teacher, nodded at a boy. 'I feel frustrated,' he said. Ms Soto-Gomez looked at him kindly: 'That's *very* good, Damien.' She asked him for an example and he thought for a bit. 'I feel mad,' he concluded, '*and* sad.' 'Mad because you don't want recess to finish?' 'Mad,' he confirmed. Ms Soto-Gomez gently coached him. 'Everybody likes recess,' she agreed, 'but when recess finishes I need to move to another place. So are we aware of how we're feeling when recess finishes? You said frustrated. What if we come back frustrated from recess?'

It was a simple exchange and a very small step. But it was character-building. More than that, it seemed to contribute to their well-being. While the idea of taking care or developing emotional intelligence could feel impossibly vague at times, Ms Soto-Gomez used specific and concrete practices. The kids were learning to identify feelings, and think about their reactions to them. Over the next eight years they'd experience hundreds, even thousands, of these teachable moments. Changing habits was hard, but for kids from inner-city Hartford, many of whom were growing up in poverty, the lessons were invaluable.

The class teacher was Ms Mullings. As a young African-American woman who'd attended the dysfunctional Hartford High – 'There were race riots going on. I was going to school in madness. I was a straight A student, but I didn't know how to deal with anything inside me' – she knew what was at stake for the kids if they didn't find a way to cope. As we chatted, she explained that they often got frustrated. When they did, she had a routine: 'OK, let's turn off the lights, let's turn on the mindfulness music, let's have a moment, let's get calm, let's think about our problems.' Recently, she'd noticed something remarkable – five- and six-year-old kids were now adopting these practices themselves. 'They go "Ms *Mullings*,"' she sang in perfect first-gradese, 'and they go right to the mindfulness corner. They take a break, relax and then, "Alright! Now I'm gonna show some *grit*." They use these words!' This breakdown to breakthrough was her favourite move. 'I love combining mindfulness with the grit right after.' And she knew her kids needed it – not only for the world beyond school, but to survive their time within the education system.

'Teaching at a community school, everything was testing, scores.' Her daughter was now at Breakthrough Magnet, but previously she'd been ranked. 'Are you a red kid, a yellow kid, a green kid?' She got depressed. This was a common theme for the teachers. For them, character education meant nothing more or less than treating the kids as humans first. The school cared first and foremost about the students' well-being. Success in the traditional academic sense was a secondary concern. 'In other schools they are looking at data. They are looking at scores and testing,' Windy had told me. 'I'm growing human beings. That's what I do.' It was the epitome of taking care. We were born not only to learn to think and to do, but also to feel. Breakthrough Magnet was going further in developing the third of these faculties than any other school I visited.

In the Mindfulness Room, kids reflected on this. The previous week the whole school had taken their MAP® (Measure of Academic Progress) tests.[22] How had it made them feel?

'When I do my test, I feel a little nervous because they give me a hard test,' offered Darrin. 'And when they give me a hard test I stop touching the mouse and I look at the teachers. And then when I feel scared I *cry*.'

'You didn't cry!' protested Ms Mullings.

'How many of you felt scared?'

Twenty-five small arms went up.

'It's a valid feeling,' said Ms Soto-Gomez.

'I felt a little nervous,' added Leyla.

'How could you finish the test?'

'I took a little mind moment. And a little brain break.'

They were just starting out, but the kids were getting it. They were vocalising thoughts and feelings that I'd never managed to bring to the surface. I wondered if some of the power of this came from creating a community atmosphere in which everyone felt comfortable sharing, in which no one would be picked on for playing a Tibetan handbell or talking about taking a brain break. I even thought it could work in the UK, if you started with kids early enough. I'd seen something like it in School 21's circle time. The psychologist Oliver James, author of *Affluenza* and an expert in depression, once told me that every teenager should be given 16 hour-long sessions of psychotherapy before graduating school. With a lesson a week combined with countless other micro-moments across up to ten years at the school, these kids were getting a lot more than that. The group wrapped up with a 'wave of gratitude', each kid Chinese-whispering their neighbour a 'fantastic transition' into recess, and I headed off to find Ms Goldstein.

'It's so much about relationships,' she said. 'Gratitude reinforces the relationship. It strengthens that spirit. It grows from there.' Ms Goldstein radiated goodness. As well as teaching her morning gym classes and running a top Magnet school, she was committed to building a better world. Prior to becoming a principal, Goldstein had spent years as a school social worker, supporting the most vulnerable kids. From the driver's seat of her SUV as we'd headed to school after the spin class, she'd explained her own theory that character development depended at heart on the daily behaviour of teachers. 'Your children love you. They know that you believe in them. They get that through actions and words,' she said. Although there was a deep science to it, this also made it hard to game, depending above all on kids experiencing a true sense of security within the school community. 'They feel very safe to take risks and take chances.'

For Goldstein well-being came before well-doing. Her approach was rooted in the work of Norman Gramerzy and Michael Meaney, who we heard about at Pen Green and who'd done so much to show the effects of poverty and trauma on the infant brain. It didn't *feel* scientific though, but joyful. Kids walked – skipped! – through the doors every morning beaming. Ms Goldstein greeted them with a big smile and kind words. They attended their weekly meditation sessions and paused for regular mindful moments in their class-rooms. The sense of community was palpable. The school put people first. Everyone *belonged*. And, in all of this, the kids were doing great academically – not as well as KIPP, but twice as well on average at reading, writing and maths as was expected of them.

The effect of the school on kids' well-being reminded me of a story I'd heard about Iceland. Two decades ago the island nation had some of the highest levels of teen substance abuse in Europe thanks to its low levels of adolescent well-being. But rather than discipline the unruly kids, the government did something different. Thanks to an insight from a US psychology professor based in Reykjavik that addictions often resulted from a natural response to stress which sought out highs, they built a 'protective' programme. It ensured all Icelandic kids had access to a natural set of highs achieved through free classes and clubs in dance, music, art and sports, rather than alcohol, cigarettes or narcotics. Today Icelandic kids were Europe's cleanest-living. 'The Icelandic model could benefit the general psychological and physical well-being of millions of kids,' urged the report.[23] Rather than coaching the individual to beat the environment, which you did with self-control, you shaped the environment to get the best out of everyone. Ms Goldstein and her team had built that culture in their school against the prevailing attitudes in the US system.

Breakthrough's community ethos fuelled kids' happiness *before* their success. Was this the lesson we should take for the future?

My last act at Breakthrough Magnet was to attend the community meeting. Earlier that morning, I'd attended a second five-thirty boot camp session. Toiling on and off steps, feebly star-jumping and failing to pump iron, I thought about Angela Duckworth and Will Smith. I *could* be outworked. I *wasn't* prepared to die on a treadmill. Was that OK? Breakthrough didn't put success first either. It wasn't trying to turn out self-disciplinarians, as they were in South Korea and I suspected KIPP might be, but kids comfortable in their skin and secure in the community around them. Last week they'd spontaneously made one of the teachers a card with 30 origami hearts when he was absent visiting his sick brother. That was gratitude, and love. In the community meeting, they celebrated the class BRICKs and followed Mariela in a whole school mindful moment before the cheer squad charged on stage to Calvin Harris's 'Let's Go' and 300 kids, adults and visitors had joined the choreographed dance.

Kids here learned to be happy, mindful community members. When tomorrow's world threatened stress, isolation and poverty, didn't it seem right to put well-being ahead of well-doing? I planned to put the question to Angela Duckworth.

We're All Gonna Die

'We're. All. Gonna. Die,' said Donald Kamentz conspiratorially, tilting forward in his chair. We sat at a conference table in the University of Pennsylvania's Positive Psychology Center. The Center was home to Character Lab, which had been set up by Angela Duckworth, Dave Levin and a private school principal named Dominic Randolph to encourage research and experimentation that put character back at the centre of learning. Donald was the executive director and had the neatly groomed appearance and kindly manner common to a new wave of male educators. Emily, Chad and Sean completed the group, each an expert in child psychology and schooling. 'We're all gonna die!' they chorused. The group maintained a steady banter and clearly loved working together. 'This is what Angela would say if she were here, literally. "We're all going to die. And don't you actually want to have done something before you do?"'

I videoed back in to talk to Duckworth a few weeks later. She explained that her mission was to change the world. It started at school.

'I worked with very poor kids,' she began, radiating a bigger sense of fun than I'd expected from the champion of grit. 'My urgent question was how we can get them to work better and harder in school.' Not because she didn't think that curiosity and gratitude are important, she explained, but 'because the achievement gap is going to grow'. Humans are wildly complex. '*I* don't have the answer,' she told me, 'and I don't think anybody has *the* answer.' But she believed we had to narrow our focus if we hoped to isolate characteristics to better understand how to grow them in kids. In *Grit*, she advocated supporting kids to find passion and purpose; perseverance would follow. 'You have to have a superordinate goal that is more important to you than any other goal,' she told me. She talked of the importance of tough love. It was universally applicable – to her own daughters as much as her Alphabet City students – and stressed the pursuit both of achievement and, above all, goodness.

It was a dog-eat-dog universe out there. Should we prepare kids for the world that's coming or for the one we wanted to see? Duckworth thought both. 'I want them to learn calculus,' she said of her own teenage daughters, 'and I want them to be able to craft a beautiful essay and I want them to be able to understand the theory of evolution.' At the same time, she wanted them to develop as good people. 'I hope they become honest and kind and grateful and really self-controlled, creative.' It felt like a noble, caring aim, but also a little hokey. Her great hope was that the Character Lab would help schools get better at cultivating those strengths in kids. The system favoured the traditional academic measures that I'd experienced at Walworth. But Duckworth knew that educators arrived at school each day to help kids grow, not to coach them through tests. Her commitment was to equip them with the psychological science that put character back on a par with academics.

'Every teacher, I think, at some level, tries to get kids to be honest, and hard-working and kind,' she said. 'I think the lack of intentionality, the lack of conscious awareness of which things work better than others, that's something I hope to personally contribute to changing.' Hence the Lab. Psychologists knew a lot about the mind. They'd developed proven practices and approaches from WOOP for goal-setting (you outlined your Wish, Outcome, Obstacle, Plan), to simple online interventions that could develop kids' belief in themselves. And they worked. David Yeager at Stanford had raised the grade point averages of at-risk high schoolers by briefly showing them online articles on the malleability of the brain.[24] Too often, though, educators were unaware of these hacks. And the reverse was true.

'I can't remember a single time that a psychologist called me and said, "Hey I'd really like to talk to an insightful teacher about what they think about this problem,"' said Duckworth. 'My experience is that teachers have brilliant insights into the minds and hearts of kids.' In the Lab, Kamentz had agreed. 'We all know that the non-academic, non-cognitive, the character is just as critical if not more essential to longer-term success,' he said. He'd been a teacher and college counsellor for almost 20 years, learning through hard experience. But now he'd seen the science and it had *opened his eyes. The Walter Mischel marshmallow task! Carol Dweck's growth mindset! The Daniel Pink research!*[25] These tools could supercharge the classroom. Bridging the gap between the science and the school was a huge part of this game. That was how doctors did it. Research and practice happened side by side. The boundaries of science were pushed within the four walls of the hospital. If it worked for bodily health, could it also work for mental health? KIPP and Breakthrough Magnet were already using techniques of cognitive behavioural therapy with real success.

It still seemed a little focused on the type of achievement that rang of gyms, diets and work. I put it to the other members of the Character Lab that it must be exhausting for them to be constantly aspiring for greatness, to be hyper-aware that any attribute of their character was malleable, could be grown. Didn't it mean you were always beating yourself up? Was it still possible to have purpose-less, meandering *fun*? Could we permit kids to? I didn't think you could design a world in which we expected all kids not only to excel academically, but also to be West Point material. We weren't Spartans.

'Does it torment me to work with a team of psychologists all the time?' laughed Chad, setting everyone else off. *Right!* The team had fun. But they'd given serious thought to the question.

'Do you have a fundamental responsibility for who you are?' he went on. That was the question. Critics of grit saw it as a pick-yourself-up-by-your-bootstraps dogma. 'You have control over what you do. Fix yourself.' But this was wrong. The Lab was sensitive in its work. 'Our interventions are mindful that people are different, that there are different circumstances.' Chad saw it in his own character. 'There are areas in my life where I am probably super-gritty and others where I am just not.' The jury was out on whether you could transfer characteristics from one domain of existence to another. If you always stuck at soccer practice, but gave up on your run around the park. The Lab's PhDs were looking into the question as we spoke. Knowing the secrets of the mind – of character – seemed a way off, but we were making progress. We knew on some level that kids who took the marshmallow in the test weren't *failing*. They were responding to an environment they'd grown up in. In some cases it really *was* better to take what you could get while you had the chance.* Though Duckworth thought it best to be tough on them, helping them learn self-control, she also believed that with love they'd 'be happier, and most importantly, they'll contribute more to the world'.

Life's Not All Puppies and Rainbows

'You might think we do this because we enjoy it,' said Jeff Li back at KIPP Infinity as an instrumental version of 'Boulevard of Broken Dreams' played soothingly from his laptop. It was my final day at the school and eighth-grade maths had kicked off with a lecture after a group of kids were benched in reading time. They seethed with quiet adolescent injustice. But teachers didn't enjoy this stuff, explained Mr Li. Not *much* anyway. 'The thing we're trying to do is not to see you as you are now, but as you'll be ten years from now,

---

* This response to the environment is particularly evident in the different outcomes of girls and boys in school and in life. In researching this book, I interviewed Laura Bates, whose brilliant Everyday Sexism blog charts instances of misogyny women are subjected to every day (and whose book *Girl Up* is a guide to how we what we can do about it). Her work shows that even within schools, girls aren't safe from sexual bullying or name-calling, with three in four young people aged 16 to 18 recently revealing they've heard girls called 'slut' or 'slag' at school, and three in five young women of 13 to 21 reporting that they've experienced harassment at school or college in the past year. When coupled with the evidence that girls generally outperform boys at all stages of education, yet go on to fall behind in earnings or occupying positions of power beyond school, this presents a troubling picture. Is it possible that the environment, which demands perfection and conscientiousness of girls and punishes them for any transgression with bullying, both propels them to higher grades and subtly undermines their confidence? The subject requires a book-length treatment, and luckily Laura has written two. Read them, they'll change your life. (Laura Bates, *Everyday Sexism*, London, Simon & Schuster, 2014; *Girl Up*, London, Simon & Schuster, 2016.)

twenty years from now. So when you don't complete your choice reading, that's what we care about.' He flicked on an image. There were a hundred senators in the US, two for each state.

'How many of them look like you?' he asked.

The eighth-graders looked around at each other. 'Twenty,' came a first guess. 'Eighteen.' 'Fifteen.' 'Thirteen.'

'Ten,' replied Mr Li, slowly making eye contact with each of the students. 'And this is the best it has ever been.'

He paused, annunciating his next words slowly. 'What we see when you don't do your homework is a perpetuation of this problem,' he added. The class was dead silent.

'At Stanford, I'd put any of you guys against my classmates.' He meant it. But it was hard. He pointed back towards the reading classroom and the teacher that had benched them.

'It's because she loves you so much. It hurts her to see you piss your future away.'

Isaiah raised his hand tentatively.

'It's a step back from your dream,' he said. Mr Li nodded.

So this was how you did *Dead Poets Society*.

'If this makes sense to you, make a small commitment to yourself,' said Mr Li. His talk had run to fifteen minutes. 'One thing I can do better. Then see by the end of the lesson, did I do it?'

The day before I'd sat agog at the back of Ms Fascilla's fifth-grade class as the young mathletes mastered fractions. Stern in manner and unapologetically exacting, she expected 100 per cent effort from the kids. 'Think. Think. Think,' she urged. 'What do you notice? What else do you notice? What else?' It was tough, and it was magnificent. Answers were considered wrong unless they were *exactly* right. Activities took place under tightly controlled routines. For paired talk, kids had two minutes *on the dot*. When hands went up, *every* hand went up. While kids solved problems individually, John Williams or Chopin played in the background. Only in Wanhangdu Road Primary school in Shanghai had I seen standards upheld to such a high level. This was how you rocketed kids from third-grade maths to high school algebra twice as fast as average – through self-control, compliance and a lot of hard thinking. Kids made two or three years' progress in her class in a year. They had to, if they were going to make it from a third-grade level to high school algebra in their four years at middle school.

But by the time they got to eighth grade, the atmosphere had changed. You couldn't expect kids to mindlessly comply any longer, and nor was that good for their future. I'd always thought the downside to self-control – the dark secret of all those two-marshmallow maestros – was that at an extreme it risked creating pliant, well-behaved drones. Ju-Ho Lee worried about South Korea's future for that very reason. But if you looked past the wall slogans, you could see that *wasn't* the character they were building at Infinity. Here it felt political. They were growing solidarity.

Mr Li taught the eighth-graders, 13-year-olds heading off to high school.

'We're trying to convince our kids to go to college,' he said after the class, 'so we could say, "Hey, go to college, it's good for you, it's good for your family, you get a better job, it's totally fun."' It worked with the youngest kids, but by the time they reached adolescence Mr Li had to 'violate their schema' of why they were in school. It was no longer about them doing the right thing, but getting fired up and becoming organised, proving the doubters wrong. Some rebelliousness was helpful if you were to defy the low expectations society held of you. Li and KIPP wanted the kids to learn a lesson about taking responsibility for their own futures. As I'd learned in Finland, they could only really do that through experience. Principal Holley explained to me that the school had recently chosen to scale back some of the intensity with which they enforced compliance in the kids. They wanted to see more joy, more personality, and that meant giving them more freedom.

'In a lot of classes,' Mr Li explained, 'it seems like the teacher's job is to win a conflict through a series of ninja moves.' He'd concluded that this worked in the short term, but failed to help kids build their own self-motivation. It was an inherently limited approach. 'If you really want to unlock student effort,' he said, 'you have to put them in the driver's seat, ensure they understand that they have agency and power.' Helping them 'understand that sense of empowerment is huge'. I saw this at work throughout KIPP Infinity.

'To give you some pieces of data,' he offered. 'Last year we had a cohort of kids about 25 per cent of whom had IEPs (Individual Educational Plans are for students with special needs). And over 90 per cent of our kids were eligible for free or reduced lunch.' These were indicators of the highest levels of challenge. Kids coming to KIPP Infinity were among the poorest in the country. Yet whereas just 6 per cent of kids in the district scored proficient or advanced on New York State's Common Core-aligned eight-grade mathematics tests, and 81 per cent of them did in the most affluent areas like Scarsdale in Westchester upstate, *92 per cent* did here. I doubted whether there were any schools in the world where kids learned so much maths in four years. Jeff didn't put it down to maths-teaching wizardry alone though.

'Yeah I teach math,' he explained. 'I'm very meticulous about the standards and the rigour. But I spend so much of my effort trying to unlock student energy through mindset, through character.' He was a lead practitioner in the Character Lab and estimated that three-quarters of his effort went into coaching kids on their attitudes. The lectures *were* a big part of it. They were legendary. He wanted kids to be aware of the choices they were making, and their consequences. On the classroom wall, two framed Nike shirts reminded them that 'Talent Ain't Enough' and 'Maybe You Should Practice', but my eye was drawn to a huge banner urging onlookers to 'Prove the doubters wrong'. This was his mantra. Mr Li had built his practice around it. He put his challenge simply. 'How do you convince a large group of people to do something they would not otherwise do?' Character development was the answer. Li cared, deeply.

As I watched his class breeze through simultaneous equations to work out the best mobile-phone tariff to buy – he was up to date, but not sufficiently so in a world of unlimited data – I reflected. It would be easy with the focus on self-control and grit to make the criticism that these kids were destined to become obedient workers, grittily sucking up all the hardship the twenty-first century might throw at them. Tempting, but also false. The Kippsters at Infinity weren't little Gordon Gekkos or Jane Fondas. They were kind, caring. A real team and family. Later on, I saw Mr Li's students giving up their free time to help others, part of an extensive peer-tutoring programme he had devised.

The idea we could WOOP ourselves to victory seemed born of scientism, offering only a partial view of who we were as learners. It risked our giving in to the forces that shaped our world – that created the poverty which many of these kids were trying to escape. Writ large, this ethic could mean an education system like Korea's – highly successful, economically productive, but leaving hidden scars on its students. What chance did kids have of finding their passion if they were trapped in an endless struggle to the top? In the last leg of my journey, I would travel on to Hong Kong to explore how kids could learn to shape a better, more caring world in which they no longer had to race. For now, taking care meant preparing them to succeed in the world they faced, and fuelling their hope of creating a better one together. Self-control and grit *could* be taught. Teachers were powerless to improve the circumstances of their students, but they could help change the effect the world had on them, reshaping the architecture of their brains and giving them strategies to cope.

The route to success is paved with the bodies of those that didn't make it. The route to happiness is lined with your friends cheering you on. I thought about Julie Goldstein and the Women's March. I thought about Angela Duckworth's tough love. Of course, it was possible to have both. You could be part Sarah Connor, body and mind honed for everlasting life. But you could also be part Sarah Connor, driven by love and humanity. In the end it was a question of values. We didn't all share the same ones, but we did have to ensure that our actions – and those of our students – aligned to the ones we held dear. Jeff Li had told me that 'children have never been very good at listening to their elders, but they have never failed to imitate them'. James Baldwin had said that. It's why you couldn't teach character from books. You had to live it.

On the wall hung a Hattori Hanzō sword, a nod to *Kill Bill*, and a visual anchor for the students as they prepared for their various exams. Below it in a display case stood a crumpled Kirkland mineral water bottle.

Nine years earlier Mr Li had taken his students on an end-of-year trip to California. In those days Kippsters didn't carry named water bottles around with them as they did now, so they ended up buying bottled water with the lunches wherever they went. The teachers urged kids to keep hold of their bottle and refill it, but of course it descended into chaos. Their coach started to fill up with plastic. Picnic spots were transformed by the end of lunch into recycling centres. There was no way of knowing who a bottle belonged to. As things fell apart Mr Li launched into one of his lectures. He talked about responsibility, community and integrity. And he told them that if they could truly demonstrate those values they would be able to hold on to their bottle for the whole trip, keeping it with them all the way home. He made them a ridiculous promise. If any one of them brought back their water bottle after they graduated college – this was ten years away for most of them – then he would personally take them out for dinner at the finest restaurant of their choosing.

Last year, he'd received a phone call. Tiara had just graduated college, and she'd remembered. Over dinner, she presented Mr Li with her crumpled water bottle. That was grit, I guess. But it also felt a lot like love.

CHAPTER 9

Mind Control

Rebels with a Cause

It is the mark of an educated mind to be able to
entertain a thought without accepting it.

Aristotle

You Call It Education, I Call It Rebellion[1]

In May 2011, while other middle schoolers were surreptitiously
playing video games instead of doing their homework, a polite,
principled, hardworking[2] Hong Kong teenager named Joshua Wong
began organising a movement to defeat the Chinese Communist
Party. Hong Kong's Legislative Committee had just announced
that all kids in the territory would have to study a new 'Moral and
National Education' curriculum that criticised democracy, aspired
to shape 'positive values and attitudes' among kids and praised the
CCP as a 'progressive, selfless and united organization'.[3] Wong saw
the policy for what it was – a Beijing plot to indoctrinate young
minds into the New China Dream. Fired up with a righteous sense
of injustice, he teamed up with a school friend, Ivan Lam, to launch
'Scholarism', an online activist group with a single aim: to put a stop
to the brainwashing curriculum. Wong was just 14 years old.

'It's not a dramatic scene,' he said when we met in a budget café
deep in the bowels of Hong Kong's financial centre.* I'd asked him

---

* In a slightly overzealous piece of self-protection, I'd avoided contacting Wong

how it had all started. After Scholarism went on to fuel a revolution, Wong had become the territory's most infamous son. His life was now filled with press interviews, political meetings and court dates. And he was understandably wary of fuelling his myth. The international media had dramatised his story as a blockbuster tale of David versus Goliath, casting the reluctant young activist as an icon of global resistance. 'The Boy Who Defied an Empire!' *Teenager vs Superpower!*[4] You could see why. With his slight build, bowl haircut and black-framed glasses, Wong was pure underdog. In keeping with the part, he was keen to underplay Scholarism's start. 'I just set up a Facebook page,' he insisted.

For months afterwards no one took the young protestors seriously. Hong Kongers were economic animals, whose gruelling ascent of the education ziggurat towards the top business courses and most lucrative careers left no room for playing politics.[5] Wong knew the feeling. When he'd once asked a primary school teacher how he could best contribute to society, she'd recommended striving to secure a job at a multinational corporation. 'Then when you are wealthy,' she explained, 'you can give donations to the poor.' His own studies included the usual suspects of 'debt and credit, market segmentation', and 'how to estimate profit'. Outside class, like most other teens, he 'just played computer games'.[6] Activism was his detour. But he had set aside his PlayStation and stuck at it.

His perseverance paid off. A few months after the online launch, Wong organised a petition, convincing 200 young volunteers to schlep out to Hong Kong's various metro stations, squares and covered walkways to collect signatures. Toting black banners depicting an adolescent brain draped in a Chinese flag, and chanting 'Defend freedom of thought! Oppose the brainwashing education!', the teams of youngsters persuaded 100,000 people to put pen to paper. Sensing echoes of the Arab Spring in the work of the

---

directly until I crossed the border into China, for fear that my email would be intercepted by CCP operatives and I'd be denied entry. As I had sat waiting for a first meeting with Wong, which he'd ultimately cancelled, three days before our eventual interview, I became convinced that the two other men in the cold-press coffee shop were secret agents prepared to watch my every move.

young activists of Scholarism, the international media converged and Wong found himself interviewed on national and global television in front of a thicket of TV microphones. Resolute in tone and a little awkward in manner, the unassuming Wong shone, with millions later watching his performance on YouTube. Political parties began to take the part of the schoolkids, joining calls to reject the new curriculum. The government, however, remained unmoved.

Momentum grew again the following spring when the CCP-backed tycoon C. Y. Leung was named Chief Executive of Hong Kong. His behind-closed-doors selection by 1,200 officials shone a light on the territory's undemocratic rule, fuelling a further surge of sentiment. When Leung then ignored a peaceful protest march and dismissed Wong's efforts in a TV interview, the leaders of Scholarism turned to direct action. That September, a month after the London Olympics and four days before the Moral Education was due to be rolled out in Hong Kong schools, a few dozen young activists occupied the main square in front of the government offices, pitching tents on the concrete floor of the plaza. Interviewed in the *South China Morning Post*, Wong quoted the Japanese writer Haruki Murakami to explain that through small efforts you could defeat great powers. 'If there is a hard, high wall and an egg breaks against it,' he said, 'no matter how right the wall or how wrong the egg, I will stand on the side of the egg.'[7]

For three days of torrential rain only a few hardy activists showed up in person to join them, most of the 200 volunteers put off by the weather, but Wong refused to give up. 'The warning light of Hong Kong's political system is flashing red,' he said, his defiant tone rallying university students and working adults into action. On the fourth day, a crowd of 4,000 people joined the protest, waving red banners proclaiming 'we are the future' and sitting through an orderly night of candlelit speeches. Others soon arrived. By day nine of the occupation, the public mood had turned decisively. A crowd of 120,000 congregated in matching black T-shirts, shutting the highways around the government offices and chanting in unison against the brainwashing curriculum. Leung's administration, surprised by the turnout and disarmed by the 'peaceful, rational and non-violent protest', capitulated. National and Moral Education was quietly shelved.

'Students can also make change in society,' said Wong in summary. I'd journeyed to hear his story because I agreed. While most kids were too busy studying and most adults too hard at work, Joshua Wong had put his values first. Seeing in education the power to shape societies and faced with a threat to freedom of thought, he had taken responsibility for a future generation. I saw it as the ultimate act of care. The slight Hong Kong college student sitting across from me defied all of the usual stereotypes about our young. I wanted to know what had led an otherwise unremarkable 14-year-old boy to face up to the world's most powerful state, and *win*. This chapter tells the story of Wong's answer to that question. It shows that we only fully realise our potential in co-operation with others, and argues that taking care means bringing up global citizens capable of building a world we wish to live in. It begins with Wong's first victory in the fight to preserve Hong Kong kids' right to think critically, to think for themselves. I knew from my own classroom just how important that was.

The Truth is Out There

The need today for superior thinking skills was brought home to me back at Walworth on the day after Barack Obama's first election, when 7X1 filed into the classroom after lunch. I'd asked, wasn't it fantastic news? Sure, sir, they'd responded uninterestedly. School rippled instead with talk of Bonfire Night, the annual celebration of Guy Fawkes's thwarted attempt to blow up Parliament. The kids were looking forward to an evening of letting off fireworks, at each other, on the estate. 'Don't worry, sir, we wear jackets!' Jonathan had reassured me, rolling his eyes. Unperturbed by their lukewarm reaction, I persevered. I'd spent hours planning a lesson around the headlines, 'Oh-bama!', 'Yes We Can', 'The Face of Change' and I wasn't going to deviate from it. Here was the first black President of the United States of America! It was a day of planetary significance.

Awaiting their awed silence, I noticed a telltale look of mischief cross Tyson's face as he slowly raised his hand. I steeled myself.

'No he's not,' he said.

A brief silence ensued. Tyson topped the detention league tables and the other 29 members of 7X1 sensed drama.

'He is,' I replied.

'No, he isn't,' said Tyson, standing up from his chair and now making his way slowly up to my desk, top button undone and shoe-laces flailing, 'I can show you.'

Opening Google at the monitor on my desk he tapped out a search, clicked a link and turned to face me in arms-crossed satisfaction.

'See!' he said, triumphantly.

The class and I watched on as a webpage flashed up onto the interactive whiteboard. 'The Seven Black Presidents Before Barack Obama'.

I took a deep breath. While in the beginning, the internet had promised to democratise knowledge, ripping out of books, convert-ing it into bits and promising to transform for ever the way that we learn, now it was causing trouble. No longer the preserve of a cultural elite, knowledge was now *for everyone* (or at the very least those with wifi). This had also thrown up some problems. Whereas previously we had shared in a single body of carefully curated knowledge, now more information was said to be produced every two days than in 200,000 years of human history to 2003.[8] A lot of it, like this site, was misinformation. But what to me was the obvious work of some autodidact conspiracy theorist blogging from a distant bedroom, to Tyson (and a growing proportion of more impressionable 7X1 pupils) had the look of truth.[9]

We now know it's not only kids from Elephant and Castle who struggle to discern fact from fiction. A 2016 Stanford University study reported that four out of five middle schoolers and most col-lege students could not distinguish between sponsored content and a real news story. 'People assume that because young people are fluent in social media they are equally perceptive about what they find there,' wrote the researchers. 'Our work shows the opposite to be true.'[10] So did my experience. The 'facts' that were now available to fit every opinion risked marooning kids (and well-educated adults) between the rock of state indoctrination and the hard place of the Facebook newsfeed.

I pulled up a news site announcing Obama as the first black President of the US. Then another. He remained unmoved.

'Distinguishing what is true from what is not true is a critical skill today,'[11] said Andreas Schleicher. It had been for some time. The philosopher and Holocaust survivor Hannah Arendt had famously written of the rise of Nazism that 'in an ever-changing, incomprehensible world the masses had reached the point where they would, at the same time, believe everything and nothing, think that everything was possible and nothing was true'.[12] In that context, our current proliferation of information promised our easy manipulation at best, an easy gateway to extremist views at worst. More than thinking critically, it seemed to me that today's kids needed to be able to think more critically than ever before. That's what had inspired me to track down Joshua Wong. If we really cared about our kids, we had to teach them to think for themselves.

Before setting off, I'd gone to speak to a man who knew more than most about the dangers of indoctrination. I wanted to ask him what our schools could do to take care of our kids in a world in which anyone might be out to lead them astray.

Radicalise This

Adam Deen was late. I sat in a nondescript foyer on London's Fleet Street staring out past the rush hour taxis at the looming spire of St Bride's Church. Strangely, the receptionist of the building hadn't heard of him or his organisation, though I knew I had the right address. Perhaps the secrecy was deliberate. Deen was a reformed Islamist who now ran Quilliam, the UK's first counter-extremism think tank. It had been founded several years earlier by Maajid Nawaz, whose autobiography, *Radical*, I clutched in my hand. I'd contacted them after two tragedies shook London schools. First, the ISIS executioner Mohammed Emwazi was found to have been a former pupil of Quintin Kynaston secondary in St John's Wood, where he was remembered as a nice, quiet boy. Then three high-achieving girls doing their GCSEs at Bethnal Green Academy had left home one day and, instead of making their way to school, boarded a flight to Syria to join the cause.

What had gone wrong? The internet was fuelling an age of extremes, reinforcing radical views, whether they were nationalist, misogynist, racist or religious extremist. I wanted to know how our schools might counteract them. The British government had launched a 'Prevent' strategy to counter extremism, but it had been heavily criticised for prompting a witch hunt (a young primary school boy who'd mispronounced his poorly drawn picture of a cucumber as a 'cucu-bum' had been reported to authorities by a teacher who thought he'd depicted a *bomb*) with little discernible upside.[13] I thought we'd have much to learn from Nawaz and Deen. Probing the roots of their radicalisation, could we understand something about how all kids could resist indoctrination?

As I waited for Deen, I leafed back through the pages of Nawaz's story. As a young British Pakistani teen growing up in Southend-on-Sea in the 1980s, he had been part of a tiny ethnic minority in a poor, white working-class area of the once celebrated seaside town. At the time, just a handful of Southend kids were from black or Asian backgrounds, and throughout his youth Nawaz was subjected to racism born of ignorance and hatred, shunned at football games by his peers and vomiting at primary school after being forced to eat sausages by a misguided dinner lady. Finding himself pushed to the edges, he first built an identity as a B-Boy when his uncle Nasir introduced him to the protest rap of N.W.A. and Public Enemy. Falling in with other black and Asian kids in the area seeking self-protection, he had a lot to protest about, as he and his friends were viciously and regularly targeted for attacks by neo-Nazi skinheads.[14]

The turn to Islamism was an easy step, allowing the young Nawaz to embrace a greater power, deeper identity and truer struggle. Following his older brother Osman, he began to attend study groups organised by Hizb ut-Tahrir, a revolutionary group then headed by Omar Bakri, whom Jon Ronson befriended and followed in *Them*.[15] There Nawaz began to see a new purpose in learning, winning a place at Newham College in East London and beginning a slow takeover of the student union with Ed Husain, author of *The Islamist* and today a fellow reformed radical. These 'Londonistan' days culminated in the violent murder of an African teenager on college grounds, after which Nawaz turned his attention full-time to

Hizb ut-Tahrir, becoming a key recruiter across Europe as he studied Arabic and Law at London's SOAS. On Hizb ut-Tahrir business in Cairo aged 24, Nawaz was arrested by the Egyptian secret police and imprisoned for four years. After Amnesty International secured his release he slowly turned his intelligence, charisma and experience building social movements to the effort of countering extremism.

Over the phone, he told me that the biggest mistake people made was to believe radicalisation resulted from ignorance.

'Education is no guarantee against extremism,' he said. Conversely, Nawaz lamented the 'brain drain' into Islamism of talented British Pakistanis during the Nineties, speaking of a 'lost generation' of people like Anjem Choudary, Ed Husain and himself, who had been leading intellectual lights at university. Nawaz also disliked the inference that Islam was a religion of stupidity. Muslim scholars were among the greatest in history, with the House of Wisdom in Baghdad the centre of learning that preserved the works of Aristotle. On the contrary, Islamists believed deeply in the power of learning. A recruiter's success rested on their ability to persuade you to use your analytical powers to see only the patterns they wanted you to.

When Deen arrived, apologising for the delay, I put it to him that schooling always included some element of this brainwashing.

'$E = mc^2$ is education, but it's not indoctrination,' he replied. In fact, he thought schools should be doing *more* to indoctrinate kids. We had to shape how they thought and felt about the world. It wasn't a simple question of mastering more knowledge and facts, he thought. 'Someone who is an extremist will be more in tune with world politics,' he said. 'The average kid won't know about any of that stuff.' Instead, he thought schools had to instil a common set of values in our kids. Deen had worked with the UK government to improve the Prevent programming, and spent a lot of time training teachers, 'the people that are in most contact with children'. He had become radicalised himself at the University of Westminster, but the process had begun internally long before, when friends of his in Southgate, North London, had become British National Party activists and turned against him.

'You feel like you're part of something bigger than yourself. That's the key thing.' Embracing extremism was a process, he explained. It had no formula, but there were indicators, such as a child feeling like and becoming an outsider. In the UK, far-right and Islamic extremism fed off each other, pushing kids out to the margins. He taught teachers to look out for signs. 'They may stop playing with their old friends, they become a recluse, they may now become angry, may be disruptive in class.' This sense of isolation or needing to belong was common to most teenagers. The brain remade itself in adolescence, becoming particularly susceptible to the influence of the peer group.[16] Nawaz found an outlet in N.W.A., gangs and Islamism, just as Aaron Swartz found it in the hacking community or Ifrah in the team and family of KSA. Others became emos or ravers, skaters or Chelsea fans, common room Marxists or Young Conservatives. It was the same impulse. We couldn't help but look for our group and with it a code.

The small fraction that channelled that search for identity into extremism often did so as a result of persecution. 'They internalise that racism and they process that in terms of an ideological battle between Islam and non-Muslims,' said Deen. Nawaz wrote about a day at school when Mr Moth, his inspiring economics teacher, had asked students to put up their hand if they were proud of the British Empire and its achievements. Nawaz, the only non-white kid in the class, was the only one that hadn't. There had been no room to debate the topic. 'It's pretty dire,' Deen continued. He saw schools squeezing out discussion and deep thinking to focus in on narrow cognitive skills. But it was clear every kid ought to be learning about ideology, whether Marxism, white nationalism or Islamism. There had to be spaces in schools to discuss these. Taking care of our kids meant ensuring they had as deep an understanding of the machinery of ideology as possible. Analysing ideology enabled you to see your own biases. You had to learn to understand the water you were swimming in, learn to look. 'The key is to get to them before *they* get to them,' he said.

I thought about my classroom and Walworth Academy. What spaces had we created? *Be the best you can be*, we'd told the kids, often. Work hard, do well in your exams. Be kind. Present yourself well. A strong community existed within the four walls. The kids felt secure there. But they identified more closely with other groups. It could be a country, like Jamaica or Turkey, Portugal or Congo, or a sports team like Millwall FC with its infamous 'firm' of hard men, or the church. Even gangs, tragically, for a few. Walworth brought everyone together. But it wasn't challenging kids to see and embrace difference. In some places it did. In Finland, where kids could remain in the same class of 30 with the same teacher for five years, each class was a close-knit family. In Shanghai, kids learned to love their country and adhere to a clear social code. In KIPP and KSA, kids lived by a clear code of 'Team and Family' and sets of prescribed values. But at Walworth we'd failed to. We had the uniform, a deeply caring group of teachers, many daily acts of kindness. But we'd existed in a bubble from the world beyond.

Over the phone Nawaz had outlined a five-point plan for school in the twenty-first century. It started with strong academics and critical thinking. Schools should also be prepared to indoctrinate kids – he didn't think the word was too strong – but into the 'right stuff', a set of shared facts and values. It was also key to avoid 'the soft bigotry of low expectations', not relaxing our norms on account of anyone's religion. There should be an anti-extremism toolkit for all teachers. Lastly, he had ideas for the curriculum. Nawaz remembered studying while he was at school the Soviet Revolution, the Second World War, Imperialism and how Britain became a democracy. But it was too distant. 'What we should really study is 9/11, the Afghan War, the modern history of jihadist terrorism.' Schools were too delicately liberal. 'We want to have that confidence,' he said. We ought to know what our values are, put it all out in the open. Talk it through. Everyone had been too scared to intervene when he was in school. 'Nobody wants to be called a racist,' he said.

In the Eighties, an American professor called E. D. Hirsch proposed that all kids should master a body of core knowledge that would give them what he called a shared 'cultural literacy'. Roughly stated, the idea was that each child should learn the things that would enable them to understand the daily front page of a newspaper like the *New York Times* or *Guardian*. Hirsch used his rule of thumb to compile a longlist of facts that every child should know. His ideas were popular in the UK with people like Toby Young, a right-wing journalist who launched the country's first free school, and schools minister Nick Gibb, who pushed for a clear, slightly old-fashioned knowledge basis to the school curriculum. Nawaz echoed his thesis, saying finally that 'you must start with the facts first of all. You need common reference points.'

It was precisely that which the web risked eroding. Everyone these days had easy access to their own facts. It was a foundation of extremist views – though of course they'd been around long before the web. Propagandists – like conspiracy theorists – made connections between certain events, facts and reference points. Our brains, pining after patterns, stories and meaning, would do the rest. Nawaz called it a 'constructed ignorance', and 'a pseudo-intellectualism that is just enough to make a teenager feel highly intelligent'. But he recognised the potency of these movements. In an age of globalisation he thought that it was inevitable that young people looking for a cause would end up at the extremes. There were the most potent brands. He called it an 'Age of Behaviour', where transnational ideas, narratives and allegiances defined people.

As I was asking Deen what schools could be doing better, he interrupted me. 'Philosophy!' he exclaimed. 'If I'd studied philosophy, I don't think I'd have been so easily indoctrinated.' He stressed the intellectual rigour of Islamism. It had been exhilarating. People thought it was all to do with coming from a broken home, failing to get an education. But it had actually been a kind of enlightenment for him. It was the first time he'd been asked really to think. Schools were leaving a gap in kids' lives and other influences were moving in to fill it. At Walworth, it meant some kids joined gangs. Deen experienced the same in North London. Extremism had been a refuge from an unthinking, uncaring world. It was only by taking that thought to an even *higher* level of rigour that he had been saved. 'Philosophy, critical thinking, the ability to scrutinise ideas, show its flaws, show its inconsistencies,' he went on. In Deen's view, philosophy *could* change the world.

'With the way the world is going now, becoming more polarised', Deen thought critical thinking was the key subject for schools. 'It's going to become essential.' His own de-radicalisation occurred later in his twenties. 'It's basically a conversation. The first thing is to befriend the person, get their trust. And then start engaging with some of their ideas. It's about unpicking.' It's a slow process. He'd gone on to establish the Deen Institute to foster critical thinking skills in young Muslims. He was adamant that all schools should be making space for this. He didn't think places like Bethnal Green Academy were making it more likely kids would become radicalised. But they probably weren't making it any less so either. 'It's not what they were doing, but perhaps more to do with what they weren't,' he said.

'We have to instil in young people universal truths and values,' Deen went on. But I thought we had to do more. Deen's best idea was a debating club. Nice, I thought, but not far-reaching enough. The problems that led some kids to extremism, to gangs, Marxism, nationalism, Islamism, couldn't only be solved in people's heads. Here modern capitalism had defeated the universal ideals of the Enlightenment. We told kids they were equal, free, could be anything they wanted to be. But in many communities it was only a story. Sometimes you turned a critical eye on the world and saw it to be *unjust*. I wasn't sure if our shared values of tolerance, hard work, family, TV and shopping could survive the inequities that were driving angry divisions in society. Levels of learning predicted votes for Brexit and Trump. But it didn't tell us all that much. Those on the liberal left were *also* more likely to be richer, and so better served by globalisation. Real critique exposed a flawed settlement. If we truly cared about our shared values, maybe we'd be educating revolutionaries.

Philosophers, even kid philosophers, only interpreted the world. The point, thought Wong and Nawaz, was to change it.

The More You Know, the More You Know You Don't Know

'Our point is not that people are ignorant,' wrote the cognitive scientists Steven Sloman and Philip Fernbach in *The Knowledge Illusion*, 'it's that people are more ignorant than they think they are. We all suffer, to a greater or lesser extent, from an illusion of understanding, an illusion that we understand how things work when in fact our understanding is meagre.'[17] This illusion had powered Tyson's misunderstanding in my classroom, just as it had fuelled Deen's

Islamism. We were prone to overestimating our intellectual abilities. In *Thinking, Fast and Slow*, the legendary Daniel Kahneman had shown that paradoxically, the less we truly knew about something, the more likely we were to perceive ourselves as experts. This cognitive bias shaped our view of the world.[18] We told ourselves a story that began first of all in our beliefs, then we went around listening to the opinions and finding the facts that fitted it.[19]

'A little learning is a dangerous thing,' went the old saying. Drink deep or don't drink at all: 'shallow draughts intoxicate the brain'.[20, 21] The internet had become a frat party of knowledge, all turbo-funnels, togas and beer-pong. Knowledge mattered in thinking, deeply affecting how you saw the world. But it was no longer clear that we could agree on a single body to adhere to, or whether it was desirable that we conferred this responsibility onto a single source. Governments couldn't necessarily be trusted. The mainstream media had been co-opted by business interests. Did that mean we had to take truth into our own hands? Sloman and Fernbach suggested that we'd always done so.

> The human mind is not like a desktop computer, designed to hold reams of information. The mind is a flexible problem solver that evolved to extract only the most useful information to guide decisions in new situations. As a consequence individuals store very little detailed information about the world in their heads. In that sense people are like bees and society a beehive: Our intelligence resides not in individual brains but in the collective mind. To function, individuals rely not only on knowledge stored within our skulls but also on knowledge stored elsewhere: in our bodies, in the environment, and especially in other people. When you put it all together, human thought is incredibly impressive. But it is a product of a community, not of any individual alone.[22]

If you really cared about thought or truth, you had to understand that it was a shared human endeavour. It was this insight – knowledge should be democratic rather than authoritarian – that had inspired a Hong Kong teenager to take on a superpower.

The Student Who Defied an Empire

In a cluttered café deep beneath the vast malls and elevated walkways of Admiralty in downtown Hong Kong, Joshua Wong looked up from his tea, marvelling at the excessive mound of ketchup sachets he'd been given with his fried breakfast and relaxing for an instant into the role of third-year university student. 'What's going on?' he asked comically. In the flat shopping centre light he sat sockless in Nike Roches and ankle-length chinos. A year earlier his trademark glasses had peered out of the cover of *Time* magazine. Wong had gone truly global, becoming the poster-boy for a new generation of freedom fighters, up there with Malala Yousafzai. 'There remains the myth of David versus Goliath,' he sighed, eyeing his egg and toast. 'The Free World Media needs to support the young generation against the Communist ghost!'

Defeating National Education had proved Wong's warm-up act. Aware that Beijing would continue to further the aims of the Chinese Communist Party via business, mass media and of course education, Wong vowed to fight on for democracy. He was a year old when the Chinese government promised self-determination to Hong Kongers within 20 years of the British handover in 1997, but it had become clear that they would not keep their word. So after building alliances with other dissident groups, Wong and Scholarism had led a new protest in September 2014, organising a territory-wide student strike followed by a surprise occupation of the civic square. Police had dispersed the students with tear gas, arrested Wong and detained him for 48 hours.

Following his release, the numbers of pro-democracy protestors grew. Buoyed by the febrile mood, a university professor called Benny Tai announced the start of 'Occupy Central': 200,000 people poured out onto the streets in solidarity, peacefully occupying the main highways at the centre of the financial district. The thousands of brightly coloured umbrellas the activists armed themselves with to protect against tear gas came to symbolise their resistance. News of the Umbrella Movement quickly spread across the world, images of the thousands of protestors standing resolute with umbrellas aloft and glowing under the streetlights were beamed worldwide.

The high school kids of Hong Kong were famous. Makeshift study centres began appearing on the streets, fuelling the romance. Even after blocking off the roads and shutting down commercial activity, the students were continuing their education. There were crates for seats, boards for desks, generator-powered lights to read by and points to plug in laptops. Teachers were on hand to help the young protestors with their work. The symbolism was clear.

I put it to Wong that this was first of all a *thinking* revolution, born in learning, rich in the power of schools.

'Education is not important in Hong Kong,' he said, in disagreement. Above us the sun shone on the crystal towers of Asia's financial centre. Workers and shoppers poured out of the MTR exit into the flat electric light of the plaza. Holding a piece of toast to his mouth he gestured to them, as if to say, *See?* Freethinking wasn't important to the average person on the street, he stressed, 'business is important'. I'd spoken to two high school students that week. When I asked their plans for the future, one had said, 'An artist – or in the art business', the other, 'An environmentalist, or in the energy business.' School had become a means to the end of university, which in turn secured your professional life.

'After you criticise it, you still need to attend school,' he added with a smile. Thanks to Scholarism he had only just scraped into the Open University, Hong Kong's lowest-ranking institution. But he had no regrets. It put him in the top 20 per cent of kids and he only wanted a degree so he could fulfil his new dream of running for public office. He was aiming for a third. Wong's real work was organising against Beijing, but he still had to play the game. In 2016 he'd disbanded Scholarism to co-found a political party, Demosistō, of which he was now Secretary-General. Its chairman, Nathan Law, had already been elected to the Legislative Council earlier that year. Wong's phone pinged constantly with WhatsApp messages from Reuters, his lawyer, other friends and allies. He wasn't your typical undergraduate.

I pushed on. Around the world there weren't a lot of 14-year-olds mobilising 100,000 of their peers. How had he become awakened to activism?

'There is a subject,' he said, pausing to take a bite of toast, 'known as Liberal Studies.'

Two years before the launch of Scholarism, the Hong Kong Bureau of Education, worried about the effect of too much rote-learning on the long-term hopes of their kids, introduced a compulsory new high school subject. Liberal Studies aimed to build critical thinking skills and broaden students' global awareness. High schoolers would study six subjects: personal development and interpersonal relationships; modern China; globalisation; energy technology and the environment; public health; and contemporary Hong Kong. It was intended as the final piece of the learning jigsaw for a school system ranked in the top three worldwide, launching a new generation of creators and entrepreneurs. The style of classes was different too, discussion-based and self-guided. In them, the kids and teachers talked technology and the environment, innovation and healthcare, democracy and the rule of law.

At the time of the Umbrella Movement every high school and university student in the territory had recently taken Liberal Studies classes. For his exam, Wong recalled memorising Martin Luther King's definition of civil disobedience: 'One has a moral responsibility to disobey unjust laws.' Perhaps schools could teach us to think. The evening before, I'd gone up to the law faculty of the territory's top-ranked University of Hong Kong to meet with a man who agreed. Benny Tai was the constitutional law professor who had launched Occupy Central, and he remained an outspoken critic of Beijing. Years before, he'd warned Wong against idealism but remained in awe of his young ally. 'He is one of the very special people in society,' he confided. We talked in his office high up on the ninth floor. Illuminated below us the multiple tree-lined levels of the university looked like an architect's model of a future urban zone. Attached to Tai's bookcase was an egg on which an artist friend had drawn his cartoon portrait. All it caught of Tai's wise, open face were his wide smile and thin-framed glasses.

'Who would you like to be,' he asked, referencing the Murakami quote, 'the egg or the wall?'

I wondered aloud if it was possible for eggs to be the types of critical thinkers that could succeed creatively in today's economy, without also learning to be the types of critical thinkers who'd turn against the wall and throwing themselves against it.

'That's what the Chinese government wants to do.' Tai smiled. It was possible that they might succeed. 'Chinese culture is very pragmatic and materialistic,' he explained. He believed Hong Kongers still prized economic livelihood over political freedom, and referenced Francis Fukuyama's famous essay on *The End of History*. In it, Fukuyama had announced an inevitable historical process dictating that once the average annual wage in a country surpassed $6,000, democracy would be sure to set in. Tai noted that Hong Kong had crossed the threshold 30 years previously in the 1980s, but democracy still didn't feel any closer. Or it hadn't until a few years ago.

'Chinese culture will still evolve in that direction,' he explained, 'but it may take longer.' Politicians sympathetic to Beijing had criticised Liberal Studies as the place 'where the opposition had planted all their opposition ideas'. An older generation felt threatened by the idea of a freethinking youth. As Mingxuan Zheng had told me in Shanghai, so Tai saw Western ideas of individual liberty and critical thinking seeping slowly into local culture. Tai looked to Socrates: 'You have to be very critical, even of your teachers.' On Facebook, critics attacked him for indoctrinating his students 'to be critical thinkers'. He chuckled at the paradox. In the university, he'd noticed a shift towards a more democratic approach. In his student days it had been very teacher-centred, but today it was different.

'I always believed that it can only go up there,' he said, lifting his hand above his head. Tai had seen learning as an inevitable historical process of human enlightenment. But he was no longer so sure. 'It *can* regress,' he said. 'People stop thinking critically because there are some more basic needs that they need to address. It's not that they like to be slaves,' he added, 'it's just that they want more security.'

It was getting tougher for people. Rising property prices were outstripping wage growth, and social safety nets were being cut. In his ground-breaking book *Capital*, the French economist Thomas Picketty showed how the grim trend would continue. His equation $r > g$ suggested that the rate of growth of returns on capital would *always* outstrip wage increases. In Hong Kong this truth was increasingly stark. The luxury malls and glittering apartments masked grinding poverty that saw three generations crammed into shoebox homes where bunk beds, hobs, fridges and clothes occupied a space the size of a double bed. Looking out past Tai at the city lights refracting softly through the *Blade Runner* rain, I thought how learning could awaken people, give them power.

The material rewards of learning were decreasing. That made it all the more important that we built our systems around an idea of care.

Back in the café Wong had finished his breakfast and was opening up. He saw Tai as a bit of an old-guard theorist and was doubtful of the role that schools could play. 'I will not learn from theory,' he explained, but from practice. It was one of the reasons for Scholarism's success. While the Occupy Movement drifted into entropy – he thought it was a talking shop, for people who enjoyed the means, rather than for reaching any target – he succeeded. 'I've not come from a traditional leftist activist group,' he explained. Instead, he learned through the meetings Scholarism hosted. Like old Greek agoras, the weekly gatherings would run to five, six or sometimes even ten hours. The students would reflect together on their momentum, evaluate the progress of the movement, outline new actions and deal with admin.

Wong's studies in Human Resource Management and Digital Marketing came in surprisingly handy.

He wanted to be clear that what had been painted as a social media revolution was turning out to be relatively low-tech.

'Social media is not for discussion,' he explained. It was for broadcasting. This was anathema to real thought. 'It's not for providing any discourse or any elaboration.' In his experience, the high schoolers deepened their knowledge through shared inquiry during those in-person discussions. A French academic, Bruno Latour, wrote that critique had reduced knowledge to rubble. Today we had to group together to build something from it.[23] Wong's meetings were creative, even Socratic, and they took place *in person*. The kids had to compromise, confront other views, take risks, think *together*. Social media was 'an effective tool for promotion', but it didn't prompt engagement. No one changed their views online. The rise of the right-wing extremism had proved to Wong that the internet didn't naturally tend towards Utopia.

Wong was a social media revolutionary who was no believer in social media, and a student who thought action a better tutor than theory.

But hadn't the Umbrella Movement been famous for its study centres? He shook his head.

'It's not because people love education,' he said. That was another myth. 'It's because too many homework.'

School in Hong Kong was as brutal as South Korea. The top super-tutors earned $4,000,000 some years, preparing the kids of the rich for a shot at the top universities. They were like pop stars, their matinée good looks beaming out of promotional posters on the MTR. The kids of the revolution cared about freedom and dem-ocracy, but that didn't mean they weren't still in a race to the top for the best jobs. It was one of the reasons for an occupy action rather than a class boycott. Students wouldn't give up their studies. The pressures of climbing the mountain weighed heavily even on the activists. Even there, he believed their victory was down to the Tiger mums.*

'Parents don't want their dominant position in their child's edu-cation to be replaced by the Chinese government,' he said. 'And it's the *Chinese* government. That's the problem.' He summed up their attitude succinctly. 'If you interfere with my life, I will say nothing. If you interfere with my son or daughter's life, I will stand up.' His own parents had fuelled his sense of injustice. He'd visited the poor with his parents when he was young, first praying for them, then, seeing that prayers changed nothing, committing himself to action rather than talk.

How did we grow more Joshuas? I looked around. In this vertigo-inducing centre of global hyper-commerce the street level had almost been forgotten. It was a non-space. Or at least it had been until the protests, when it had come alive with people. Learning was our most human endeavour. We did it best together, not alone. It was hard, took meetings, talking, sharing, time, everything that we were forsaking in our era of brainstem-probing content and online

---

* In 2011, Amy Chua, a law professor at Yale, caused a global sensation with her book *Battle Hymn of the Tiger Mother* (London, Bloomsbury, 2011), a paean to tough-love parenting. No one gets in the way of a Tiger Mother, not even Xi Jinping.

chatter. Haruki Murakami's eggs and walls echoed in my mind. 'We must not let the system control us, create who we are,' he had said. 'It is we who created the system.'[24]

It meant developing communities of freethinkers. Not far from the scene of Maajid Nawaz's long march through the institutions was a school attempting to do just that.

The Land of Us and Them

After projecting an image of curry and rice onto the interactive whiteboard, Mr Hartley turned to the 26 Year 2s sitting expectantly in an irregular circle on the carpet. A long line of masking tape stretched from one side to the other, labelled 'most English' at one end and 'least English' at the other. The six-year-olds had just discussed whether Mo Farah, Team GB's Olympic hero, was English, Somalian or something else ('He comes from Jamaica, I think,' said Riley, whose parents were from there), decided Englishness couldn't be defined only by manners ('They open doors for grandmothers in other countries too,' Igors argued) and agreed that nationality had something to do with how you talk ('My friend's Lithuanian,' said Daisy, 'and she doesn't speak good English'). Next up was our nation's favourite dish.

'Is curry and rice more or less English than Mo Farah, manners and the Queen?' asked Mr Hartley.

In their turquoise sweaters and black trousers, the budding philosophers appeared deep in thought.

I'd come to see them out at Gallions Primary School in Beckton, East London, because I'd heard that these classes were showing that *all* kids could learn to think. The local area was famous for its old gasworks, which had stood in for war-torn Vietnam in Stanley Kubrick's *Full Metal Jacket* and today it still struggled, out in a distant corner of Newham, one of the UK's most diverse boroughs and also home to School 21. The intake at Gallions was mixed. Of 670 kids aged three to 11, about one in five had special educational needs, one in three had been eligible at some point for free school meals and three in five spoke English as a second language.

Mr Hartley chose a hand. Keisha stood up.

'Curry and rice is least English,' she began. *Why?* 'Christians don't really eat that,' she continued. 'Muslims eat that.'

Mr Hartley continued, completely unruffled. Many teachers would have shut her down, just like I'd tried to with Tyson. But he let the scenario play out. Did Keisha know any English people that were Muslims?

She thought about it. She did.

Mr Hartley opened it up to the whole group. 'Has anyone here ever eaten curry and rice?' he asked.

Twenty-six hands went up.

Keisha reflected for a moment, then retrieved the 'curry and rice' label from the masked line and moved it a little closer to 'most English'.

'Do you agree or disagree?' asked Mr Hartley, holding his chin with his thumb and forefinger, miming thinking. A boy in black Adidas tracksuit bottoms took a turn. He nudged the label still further towards 'Most English'.

Mr Hartley asked why he had made that choice. 'A lot of people eat it in England,' he said.

When I'd travelled out that morning on the DLR to the low-rise estate where the school was based, I hadn't expected to find Beckton's own Socrates. With its many car parks and occasional trees, it hadn't been much to look at from outside. But within the walls was an oasis of prize-winning gardens, lovingly tended vegetable patches and outdoor play areas teaming with happy kids. And Mr Hartley was a master. No child was corrected or told what to think. Instead he probed the group with careful questions. His class was part of a programme started in the Seventies based on the ideas of the American education theorist John Dewey who said kids should learn through a method of 'community inquiry', a democratic approach to deepening knowledge and understanding. It was known as P4C, or 'Philosophy for Children'.

In their circle on the carpet, the 26 trainee Platos transitioned into a new scenario, shaking off their old identities with full-body wiggles. They were going to journey to a faraway land.

'This,' announced Mr Hartley, 'is Usia,' pronouncing is us-ee-ya. He pointed to the westernmost of two green islands on a blue background projected on the whiteboard. 'And this,' he continued, indicating the east, 'is Themia. You are all from Usia, which means you are *Users*.'

The Users nodded in solemn agreement as Mr Hartley began a series of thought experiments. Imagine that you fly to Themia on holiday, he began. Are you now a Themian? The kids looked around uncertainly. 'Yes,' suggested one. 'I think it's both because you went to both countries,' said another. 'No,' added a third, 'you're Usian because you speak that language.'

'What if you were to go for *three* weeks?' asked Mr Hartley, hamming it up. 'You're still Usian because you live there,' came a reply. What about for *one year*? 'Just because you go there for one year and learn the language,' said Rhinus, who had come to the UK from Poland, 'you're still an Usian.' Why? 'Because you were *born* there!' he sing-songed out to the group. Mr Hartley now flipped through a series of images. Locusts. Earthquake. Fire. While the Users were on holiday in Themia, Usia had been destroyed. There were no homes, no food, no way of ever living on Usia again. The Users fell silent. 'Are you still Usians?' asked Mr Hartley.

'I think we're Themians,' said Christophus.

'You were born in Usia. You're not a Themian,' said Rhinus.

'Just because you can't go back doesn't mean you're from that country,' added Daisy. But it was no longer so clear.

Mr Hartley introduced a new problem. He had found out that Themians didn't like *fruit or manners*! These were the lifeblood of User (and Year 2) culture. What were the Users to do? Should they continue to eat fruit and use their manners, or should they give up their fruit and manners in order to fit in with the Themians?

'If you don't eat fruit you could get sick by not being healthy,' said Dana. 'Then your bones could get rotten.'

'You could just tell people that you're from Usia and that you like fruit and that it's healthy,' offered Marius.

Christophus had a more troubling thought.

'We would stop eating fruit,' he said. He was worried about being singled out and bullied. 'If someone sees us eating the fruit then they are going to tell all of the people, then they are going to come and ask so many questions.'

I was struck by the depth of the kids' thinking and the openness with which they shared opinions. In an angry and polarised world, Mr Hartley had made a space where kids could talk it out together. Lisa Naylor, the P4C lead at Gallions, told me that a youngster in a North London primary school had told her group that 'all Muslims are terrorists'. The deputy head had shamed the girl, telling her it was disgusting to say such a thing and that she should never repeat those words. It was the wrong response. 'Shutting it down is really dangerous,' said Naylor. 'Where does that view then go?' Online, I guessed. The attraction with anonymous online communities like 4Chan was that you were free in those spaces to share any opinion that you held, however far beyond the pale. If we didn't create safe offline spaces in which to voice these thoughts, then they'd never be heard and challenged.

'You can't be a different religion because you have to stay that religion,' piped up Keisha. 'If you change your religion God will be sad.'

We had to become conscious of the stories that we told ourselves, to understand what we believed and why. In order to escape the reams of facts that could fit any tale we wanted to believe, we had to create safe, shared spaces in which we could voice and evolve our beliefs. Taking care of our kids meant doing so within our schools. Adam Deen had said philosophy and debate were our first defence against extremism, as they were against any form of indoctrination. Schools were the place where we could bring these ideas out in the open, share and discuss them in person. At least they should be. Joshua Wong had said that much of his education had happened on the street.

'Maybe *they* can learn to eat fruit,' said Riley, thinking hopefully of the Themians. 'Maybe I can use my manners and they will learn to use them too.'

A Little Knowledge is a Dangerous Thing

Critical thinking didn't simply mean mastering a common set of facts. Nor was it only a question of internalising the content of the culture in which you were immersed. Instead, true learning meant accepting that in some spaces there were no answers except those we created among ourselves. If we didn't, others would do it for us, and not only in China. Democratic governments also sought to shape the way their citizens thought. In *Propaganda,* Edward Bernays, founding father of Public Relations and nephew of Sigmund Freud, wrote that democracies were untrustworthy but could be manipulated through the 'manufacture of consent'. After cutting his teeth as a propagandist convincing the American people to support the effort of 'bringing democracy to all of Europe' during the First World War, Bernays became the field's first expert. 'The conscious and intelligent manipulation of the organised habits and opinions of the masses is an important element in democratic society,' he wrote. 'Those who manipulate this unseen mechanism of society constitute an invisible government which is the true ruling power of our country.' He saw schools and textbooks as a crucial channel of influence,[25] fuelling the opinion of intellectuals like Noam Chomsky that even a democratic country isn't free from indoctrination.[26]

Today it's not only the intentions of governments, but businesses that should concern us. The writer Duff McDonald has shown how corporations influence the content of case studies at the prestigious Harvard Business School, whose alumni go on to shape the world. An academic and tech expert, Tim Wu, writes that in California cash-strapped schools have put kids at the mercy of advertisers.[27] His book *The Attention Merchants* begins with the sorry tale of the Twin Rivers Unified School District in central California, which in 2012 partnered with EFP, the Education Funding Partners, to 'open the schoolhouse doors' to corporate advertising. 'We need to be innovative about the assets we have and learn how to bring in more revenue,' said a school district spokesperson, indicating Twin Rivers' desire to think creatively about how it could sell a portion of its students' attention to big business which knows only too well that children's minds are the most open to building lifelong brand loyalty. Another school board in Florida sold the branding of its report cards to McDonald's.[28]

If we really cared for them, we had to ensure our kids learned to think critically, and we had to free them from manipulation. Yes, they needed grounding in a set of shared facts and cultural sources, but the problem remained of working out what facts and who ought to decide. Should it be the Chinese government? The students and parents of Hong Kong? A newspaper editor? And what of the internet? The web was awash with echo chambers and conspiracy theories. It was a place where everyone wanted to seize your attention and put it to their own ends. What Tim Berners-Lee had made for everyone was now occupied everywhere by someone. 'The internet is the first thing that humanity has built that humanity doesn't understand, the largest experiment in anarchy that we have ever had,' said Google's CEO, Eric Schmidt.

Joshua Wong had insisted that his was not a social media revolution. The most vital learning experiences demanded open-mindedness, embodied discussion, ten-hour Monday meetings. They recognised that although facts were sacred, truths were often made by people, agreed and arrived at through inquiry and discussion. Our schools had to become the stewards of those debates, aiming to grow both students able to argue and analyse, and citizens practised in compromise and politics. Teachers like Mr Hartley and Ms Naylor doing so. I'd once heard someone define education as the capacity not to be cheated.[29] It was important that we weren't hoodwinked, but then we couldn't sink into *trust no one* conspiracy theoryism either. We needed shared values and facts. But we had to arrive at them through democratic discussion rather than indoctrination. Learning to think meant knowing something, but above all it meant learning to *doubt*.

In his essay *The Braindead Megaphone*, the American short-story writer and MacArthur Foundation million-dollar Genius[30] George Saunders imagined a world in which people walked into rooms hollering opinions at the top of their lungs. He imagined, in a sense, our world today. There were, he wrote, two types of people, the Toms, after Sawyer, and the Hucks, after Finn.

> The United States of Tom looks at misery and says: Hey, I didn't do it. It looks at inequity and says: All my life I busted my butt to get where I am, so don't come crying to me. Tom likes kings, codified nobility, unquestioned privilege. Huck likes people, fair play, spreading the truck around. Whereas

Tom knows, Huck wonders. Whereas Huck hopes, Tom presumes. Whereas Huck cares, Tom denies. These two parts of the American Psyche have been at war since the beginning of the nation, and come to think of it, these two parts of the World Psyche have been at war since the beginning of the world, and the hope of the nation and of the world is to embrace the Huck part and send the Tom part back up the river, where it belongs.[31]

Our school systems these days were turning out Toms. We were increasing kids' cognitive abilities, but forcing them into a bitter race to the top resulting in an epidemic of depression. Really, we should be bringing up Hucks, raising kids that cared. Benny Tai had it right. We had to teach our young to hold two conflicting ideas in their heads. On the one hand, there was expertise and knowledge. Your teacher *did* know more than you about some things. Most sources could be trusted. It was worth trying hard in your exams. On the other, you had to learn to question what your teacher told you, think critically about what you read online and in books. And you had to question your own opinions and beliefs. That was the legacy of Daniel Kahneman. The greatest gift of Philosophy for Children was that it got kids thinking about thinking. Metacognition, the academic term for this, not only boosted kids' grades,[32] but it made them a little more like Huck and a little less like Tom.

'Instead of just focusing on GPA, why don't we care about the city that we live in and the city that we love,' Wong had asked me.[33] We had to build. Saunders put it best: 'Don't be afraid to be confused,' he wrote. This attitude led people to think together. It bred tolerance, fuelled connections. To have all of the answers was to deny some fundamental part of what it meant to be human. If instead there was always something to learn, then there was always a reason to reach out. If we truly cared about our kids, our schools would take *this* most seriously. 'Try to remain permanently confused,' continued Saunders, 'anything is possible. Stay open, forever, so open it hurts, and then open up some more, until the day you die, world without end, amen.'[34] Amen.

PART IV

CODA

acing exams. Throughout my travels, I'd seen schools cultivating excellence in all areas of human learning, thinking anew about the potential of our human minds, doing better at fostering creativity and taking care to ensure the success and wellbeing of everyone. As a last step in my journey, I wanted to find a school or system that cultivated the full range of these learning faculties. Was there a way to mass-produce an individualised approach to education that meant all kids fulfilled their potential? If I could find the answer anywhere, I suspected it would be here, where the winds of digitisation, automation and global connectivity blew out across the world from a small strip of Pacific coastline, and the Kool Aid had seeped into the water supply.

I had no complaints either about the 365-days-a-year sunshine that the denizens of San Diego were waking up to.

'The weather is like this *every fucking day*,' grinned Larry Rosenstock a couple of hours later, gesturing at the streaming fall light and pausing on each word. We sat at a round table in his office, framed photos of exotic travels on the shelves behind him, the mug between us bearing his motto: 'Cultivating Creative Noncompliance'. He was a fast-talking, no-bullshit iconoclast. *Larry David meets Werner Herzog*, I jotted in my notebook. 'Since 1977 only two weather events have got in the way of Sunday golf,' he told me. In neon-laced Asics and a grey long-sleeved top, he might have come straight from the course. After a childhood studying at stuffy US prep schools, he'd worked hard all his life to dodge suit-wearing, just as he'd worked hard to rethink schools so that others didn't suffer his fate. Rosenstock was CEO and founder of High Tech High, a Southern Californian charter school network launched at the turn of the new millennium with the goal of ensuring the next generation of the area's kids were equipped to succeed in tomorrow's world.

I'd come to visit because it was working. Though entry was by zip code lottery, meaning the school had a diverse intake, *every* student was making it to college. A quarter of those went on to study science and technology degrees, far above average proportions. I'd also heard that it was a modern-day Utopia that was achieving these results less through dazzling technology, more by unlocking potential, embracing craft and building community. Here in San Diego's sprawling suburb of neo-haciendas and mock-adobe ramparts there supposedly nestled something real. High Tech High was a freethinking, communitarian zone for kids that was growing out and altering the surrounding city. It was housed in a series of old military training facilities, purchased from the US Navy. In the distance, a vast star-spangled banner rippled in the wind above the naval base, home to the Pacific Fleet.

'I have two imperatives,' added Rosenstock. 'The first is that we educate these kids to the best of our ability. The second is to change the world.' He meant it. 'Hopefully some of these kids will.'

Earlier I'd gone on a tour with a twelfth-grader, Coleman. Dressed in Uniqlo camo HeatTech with a flat-top haircut and diamante stud in each ear he was typical of the students ambling between classes, who all resembled characters in the movie *Hackers*. Coleman wasn't particularly interested in tech though. He was graduating that year and hoped to attend culinary school in New York. Cupcakes were his speciality, and his dream was to set up his own bakery. The school was built on principles of inquiry and craft, of learning by doing, and kids were encouraged to cultivate interests and engage in practical pursuits like these. Before opening High Tech High, Rosenstock had worked for the US Department of Education studying top schools for the New Urban High School project.[1] Three of the factors they'd identified for future success lay at the heart of the High Tech High mission: personalisation, adult world connection and common intellectual mission.[2] Education here meant learning to think, to do and to care.

Coleman and I stopped in a vaulting atrium at the Vietnam War project. Ninth-graders had ventured out into the local community to interview veterans, collating a polyphonic oral history of the war that was displayed at a local museum. Nearby a repurposed cigarette machine vended $5 student artworks. As we admired it, I spied a group of kids launching rockets hundreds of feet into the air out on the school field. Half of their time was spent on these cross-curricular, real-world projects, and they looked like fun. Coleman's class was in the midst of 'Coding Structures, Decoding Identities', which spanned humanities (where he was writing an essay on photography, particularly the work of a Brazilian artist making installations from trash and recording them), calculus (where he was exploring the golden ratio in plastic surgery and plants) and multimedia (he was building and coding a computer-powered spectral display). The other 50 per cent of class time was used to prep kids for their college entrance tests. Two quotes on the wall captured the learning philosophy.

> The most exciting phrase to hear in science isn't 'eureka!' but 'that's funny'.
>
> Isaac Asimov[3]

> Success is stumbling from failure to failure with no loss of enthusiasm.
>
> Winston Churchill[4]

Coleman joined his class and left me alone to roam around, a freedom unheard of in schools. In most places I was accompanied by a teacher, even at times by a troop of local authority workers, officials, lecturers and trainers. In my own classroom, though I hadn't known what I was doing, only four other adults had ever come in to observe me. Schools were usually closed, furtive systems, lacking in transparency, as though we were ashamed of teaching. High Tech High had flipped this assumption on its head. Here teaching was an open profession. You were asked *not* to knock on the door before entering – it might disrupt the flow of learning. Instead, you entered quietly and found yourself a spot to sit and observe. The teachers and kids were as accustomed to the presence of unknown adults as doctors and patients were in teaching hospitals. Five thousand people passed through the buildings each year. That was how you learned.

'We're basically open source,' explained Rosenstock. It appeared to be as true for students as it was for visitors.

In a seventh-grade classroom, a Vietnamese-American girl was giving feedback to a white surf-bro boy with long blonde hair, spiking it with *dudes!* and *woahs!*. An African-American student in a blue High Tech High hoodie explained that they'd just completed a group ritual known as 'the Walk'. When seventh-graders started at High Tech High aged 12, all 120 of them were bussed out to Border Field State Park. There they took in the tall iron railings running down the beach to the sea that divided the US from Mexico, before beginning a three-day, 23-mile trek back to school. Together the seventh-graders and their teachers camped out overnight, rooted themselves in their local community and deepened their sense of where they were from. Teachers had designed a project around the walk in which kids would write comic books based on 'dualities' they'd seen on the way. They hoped that through noticing extremes of home ownership versus homelessness or technology versus nature, the seventh-graders would begin to develop their citizenship.

Woah-girl had given surf-bro two *wows* for his first draft (colourful, organised) and one wish (spellcheck!).

Just a few days into the new school year, they were already developing a strong sense of community.

'I don't watch Justin, period,' said blue-hoodie, eavesdropping on a conversation about Bieber. He was struggling with his laptop.

'*That* is not a citation,' said woah-girl, continuing to review surf-bro's outline. They had the easy co-operative manner of the youngest kids, or most skilled adults.

'You are *so* far behind,' joined in a girl in flip-flops, 'that was like two weeks ago.' No use holding back.

'Why are you so useless, computer?' groaned blue-hoodie, banging his head down on the table for effect.

The humanities teacher noticed he was struggling and stepped in. Was he doing OK? What had he typed into Google? The questions she used put the onus on him to think carefully through his own challenge. 'Maybe we need to add another word. What is another word for caring?' He couldn't think of one. What did we do if we wanted to find out a synonym? He started tapping on the keys. 'That's right, a thesaurus. Once that pops up, type in "care" and see what list comes up.' I thought about Britain's Brightest Student Daisy Christodoulou's insistence that we continue to learn knowledge. I was sure she was right that it was vital for developing our thinking skills. And yet these kids *would* still always use Google, or its descendants. Blue-hoodie was learning how.

'We live our life with technology,' sighed a slight boy in a grey T-shirt. He was nostalgic for the mountains at the border.

On top of stories in humanities, the walk fuelled a project on population mapping and demographics in calculus, while in the Makers class next door small teams of kids were using CAD software to render three-dimensional topographical maps of the region that they'd hiked through. Teams had been assigned different quadrants, which they'd render virtually, then laser-cut in plywood before assembling the component pieces jigsaw-like into a model about the size of a backyard swimming pool. The teacher's lesson introduction felt like a creative briefing at a design agency. After a girl with white-blonde hair asked if the pieces wouldn't fall through each other when they were laser-cut, he explained that for each elevation they'd have to design a separate layer for each contour, so that they could be stacked one on top of the other.

'Ohh!' she said.

'That's what learning feels like,' he replied, 'when you go "Ohh!"'

'Or "Great Scott"!' said a boy in *Back to the Future* fancy dress holding a home-made hoverboard.

'Yes, or "Great Scott"!' answered the teacher, smiling.

Set-up over, they got down to it, to shouts of 'Group meeting!' and 'Meeting of the group!' It was still thrilling to see the ease with which the kids, whose families hailed from every continent, pitched in together. They were already doing the type of work for which someone might hire them, so high were the standards. Thinking of my own bookish education, I felt a pang of regret. The kids *loved* it here, learning to think, to do and to care. Couldn't all schools look like this? It would mean improving early-childhood learning and boosting primary school provision. If we could, I was convinced 99 per cent of kids could be brilliant readers, writers and mathematicians by age 12 (compared to around half in developed countries and just a fifth in the developing world today). If more schools managed that, they really *could* change the world.

'The world is changing and schools are not,' said Rosenstock. The question, he thought, was whether we had to drive a learning revolution from the ground up, or if 'the ways that the world is changing will be an impetus enough'.

He wasn't too hopeful about the latter. Learning was the ultimate long game, but the political cycle militated against the visionary thinking and co-operation needed to build consensus and work together to achieve it. 'We have a new President every four years. For two years before the election, the press is all about who is going to be the next President. That means I am spending *half my life* on this planet surrounded by media worrying about who is going to be the next President.' The truly meaningful act for the future was to take matters into your own hands, owning the development of a new generation of kids and having *them* change the world. Angela Davis, the great American civil-rights activist, said that we fought today not for our own future, but for that of our grandchildren.[5] With its focus on inquiry, community and co-operation, High Tech High was trying to plant the seeds for the version of the world we needed. It was inspired by thinkers of the past, like Socrates.

'Knowledge is not something that individuals do by themselves,' said Rosenstock, 'but something that we do together.'

For him, learning began with acquiring an understanding of the stories and knowledge that we held in common, as E. D. Hirsch proposed. Beyond that we had to learn to work together to apply that knowledge. Schools had to guide us in both of these aims. 'You could teach some good classes, get some good scores, answer well to the kids,' he continued, 'but that's not enough.' He thought schools threw up dilemmas all the time to help kids learn what it meant to live together. 'Imagine two second-graders studying the Civil War,' he said. 'A white kid turns to a black kid and says, "It's a good thing it turned out that way, otherwise I'd own you." You're a second-grade teacher. What do you do? Or in third grade a girl says, "I have two mothers."' Rosenstock had once been visited by mothers of two transgender kids who identified as girls. They were concerned about them using the mixed bathrooms and wanted them to have a private one. Worried it would stigmatise them, he gave the girls the option to use the private bathroom if they felt uncomfortable. '*They never did*,' he added. 'The girls didn't care.'

It was complicated. There were things we were better off simply teaching kids. How to read and write. The laws of maths. The use of a laser cutter. Facts from our shared history. It would be a waste of everyone's time if each generation had to 'discover' these things for themselves.[6] That was the point of culture and Michael Tomasello's cultural ratchet effect. But, afterwards, there seemed to be a lot of other things that kids needed to wrestle with. *What* to read and write. *Where* to apply maths. How to *interpret* history. *The ends* to which we should apply our latest technology. The best way to *live*. These were places where knowledge wasn't set in stone, where it would never be. We had to begin debating these dilemmas in school. Memorisation grew cognitive capabilities, but the *medium* of that learning communicated an implicit message of a fixed and certain world. High Tech High students were preparing for uncertainty.

'We have only two rules here,' laughed Rosenstock. The first was 'no roller-skating', the second no talking to journalists. He didn't want kids to do things *because they were told*. He wanted them to take responsibility, to question, like the kids in Finland or at School 21. There were no bells, no rules, no public address system. The kids owned the place, and they felt like they did. 'Peer pressure is underappreciated. It's not just a negative influence, it's a positive influence.' The kids didn't receive grades either. Projects were given a one or a zero. No data was held on how kids were doing, what background they were from, whether they were behind or in front. Teachers and students looked out for each other instead. Yet *everyone* made it to college. Rosenstock resisted a business model – 'the ridiculous English, History, Math, Science thing which happened in 1896 by industrialists to standardise learning' – in favour of an earlier form of human organisation, the community.

The corporate model that had fuelled an explosion in literacy had also brought competition, stress and a narrowing of our aims. Steven Pinker argued that we ought simply to aspire to increasing the sum knowledge of humanity. Societies with the highest levels of learning measured in literacy, numeracy and university graduation were the healthiest, wealthiest and most peaceable. He was right on one level, but it was also true that previous generations had sleepwalked into crises by waiting too long to respond to the signals that change was needed. There were plenty of highly educated people working at banks or in governments to promote their own interests above those of everyone else, or, in the case of oil companies and climate change, against the long-term interests of the world. Learning, like technology, was not only a force for good, but could be used for exploitation.

Thinking anew, doing better and taking care meant building a more human model of schooling, as Rosenstock was doing at High Tech High. Although it was still early days for our understanding of systems, I'd heard of some research that suggested how we might. If we were natural born learners, our ability to teach and co-operate honed through hundreds of thousands of years of society, wouldn't it be better if rather than changing ourselves to become part of the system, we fit the system around us?

Getting More Out Than You Put In

In 1993 a British physicist at a loose end had a similar hunch about our cities. Having just seen funding cut for the Desertron, a multibillion-dollar, 50-mile-long particle accelerator planned for Waxahachie, Texas, Geoffrey West sensed which way the wind was blowing and turned his attention from particle physics to people. Urbanisation was rapidly accelerating worldwide and it was predicted that by 2050 three-quarters of the planet's 11 billion inhabitants would be crammed into ever-denser metropolitan zones. 'We spend all this time thinking about cities in terms of their local details, their restaurants and museums and weather,' he told the *New York Times*.[7] Instead, his analytical mind, honed at Stanford University and the Los Alamos National Laboratory, saw the potential for something more. Could it be that every city was shaped by a set of hidden laws?

Working with fellow lapsed physicist Luís Bettencourt at the Santa Fe Institute, a theoretical research lab in the Texas borderlands, West began his research by looking at natural systems. He wanted to understand the laws that governed the scaling of biological life, or 'how systems respond when their sizes change'. Despite the incredible diversity and complexity of nature, West noticed that all living organisms grew 'in a remarkably simple and systematic fashion across an immense range, from cells to ecosystems'. It led him to propose a *power law*: doubling the size of an organism didn't mean doubling the energy needed to sustain it, but increasing it by only 75 per cent. Put simply, the larger and more complex a living system became *the more energy efficient it was*.[8] Economies of scale were a fact of the natural world. Could the same hold for our human systems?

The question led West to cities, the appearance of which resembled organisms, their vast networks of streets reminiscent of the veins in leaves or our central nervous system. Armed with a set of questions – When cities got bigger, what did it mean for productivity? If a company grew, did profits increase at the same rate? – he went in search of data to test his suspicion that cities functioned like living things, trawling the internet, libraries and dusty never-before-read reports on the demographics of mid-size Chinese metropolises or the intricacies of German regional infrastructure. Two years later he and Bettencourt had their answer. Cities *did* appear to be governed by laws. In fact, those laws were so predictable that knowing only the population of a city – 200,000, say, or 2 million – the researchers were able to predict with 85 per cent accuracy the average income of its inhabitants, the surface area of its roads and even the speed at which people walked on the pavements. It didn't matter if the city was in Sichuan, Scotland or South Carolina.[9]

West and Bettencourt's law said that when a city doubled in size, its productivity increased by 15 per cent per capita, meaning every citizen became that bit wealthier, more productive and more creative on average. 'This remarkable equation is why people move to the big city,' he explained. 'You can take the same person and if you just move them to a city that's twice as big, then all of a sudden they do 15 per cent more of everything that we can measure.'[10] Things like production of patents per population also increased according to the same law, although on the flipside so did crime and disease. Moreover, the energy, infrastructure and resources that bigger cities needed to achieve this extra output decreased by 15 per cent. Cities were organic, leaderless, relatively unplanned and barely managed, but they still achieved brilliant economies of scale. Where nature's power law said you slowed as you grew – giraffes strolled, mice scurried – cities became faster. The reason, thought West, was *social*. Cities connected people. Interaction and co-operation powered learning and innovation. Like High Tech High, they were open source.

'It's the freedom of the city that keeps it alive,' he said.[11] Buoyed by their discovery, West and Bettencourt decided to look at other human systems, beginning with corporations. Yet as they examined data for 23,000 companies, they made a startling discovery. As corporations grew in size, they were subject to the inverse of the city effect. Profit per employee actually *decreased*. West hypothesised that when companies scaled, so did their bureaucracies, stifling efficiency rather than fuelling it. Worse, companies frequently *died*, with a typical lifespan of just 40–50 years. He suspected this was due to the fact that when companies were on a downturn, the first thing they cut was research and development. 'This kind of thinking kills them,' he wrote. Whereas cities were more organic and therefore more productive, businesses petered out because artificially imposed structures put an end to learning.

Reading West's research, I was struck by its implications for our schools. Advances made in learning in the twentieth century *had* rested on a corporate model. For decades governments borrowed from the corporate sector to improve the way that we learn, obsessing over targets, introducing accountability measures and becoming slaves to data. The adoption of this model coincided with real gains in the basics of literacy and numeracy worldwide. And yet West's research seemed to suggest that paradigm may also be limiting how – and how much – our kids learn. Perhaps management theory, for all that it improved test scores, was also stifling the learning, doing, caring systems we needed.[12] Could we imagine a system instead that supported our natural ability to make the most of human community? If we thought anew, did better and took greater care, we could create an approach that amplified human systems, rather than undermined them.

In Boston, I'd gone to see Mike Goldstein. He was the founder of Match Education, a wildly successful charter school and teacher-training network in Massachusetts. He was on a year's sabbatical, his LinkedIn profile reading 'Amateur Dad' and his living room decked out like an adventure playground. Before his break he'd spent three years working as Chief Academic Officer for Bridge International, a network of 500 $5-a-month private primary and nursery schools across Kenya, Uganda, Liberia, Nigeria and India now educating 250,000 kids. Bridge had aimed at perfecting the business-type model that was succeeding so well in delivering literacy gains, and it was working for kids in the schools, who – at low cost and significant scale – performed better in the basics of reading, writing and maths than they would have in other schools in those countries.[13] But critics had also accused them of overloading on technology, with teachers reading out scripted content from preloaded tablets, and kids using similar devices to complete many of their tasks, the closest we'd got to robot teachers.[14]

You had to take the world as it was, explained Goldstein. Schools like High Tech High or Hiidenkiven worked because they had access to great teachers. But using only exceptional people meant you couldn't easily scale those schools. Bridge accepted this fact. In Kenya, the number of people who'd so far received a good education was limited. There wasn't an endless supply of talent. You had to compromise. 'If you're building to scale,' he said, 'make sure you have systems that work for the median teachers you think you can get at scale.' In Atul Gawande's now famous essay on the Cheesecake Factory, he looked at what it might take to achieve consistent high quality across all of the outlets of an upmarket national restaurant chain, and whether this could teach the health sector anything.[15] Goldstein applied the same line of inquiry to schools. 'Are you designing to have one really high-end restaurant, or are you designing something to take the labour market as it currently exists and give the customers a good experience?' he asked.

Goldstein thought there was a place for both. 'I'm glad there exists something called a Starbucks,' he said. 'And I'm glad there exists something called a Mom-and-Pop independent coffee shop with a twist.' But for me, the problem with Starbucks education was that the people who made up the system – the teachers and school leaders – were a means to the end of student learning, playing the role of cogs in a learning machine. It implied that our place in a hierarchy was settled, our human abilities limited, and that the best use of them was to follow a single, repetitive work routine. Wasn't this the kind of thinking which meant our school models hadn't changed for a century and a half? The Mom-and-Pop coffee shop model was different. It was built on an idea of craft. You could experiment, deviate from the menu. Employees might learn about coffee, cooking, or customer service, or even become partners in the business. But it did have its own risks. For every one that thrummed with brunch-time hipsters another was boarding up and eBaying the cold press.

It struck me that so far I'd almost seen two distinct missions in the schools I'd visited. KSA and KIPP worked with kids that were starting far behind, and so focused on social mobility, systematising an approach to amping up achievement and making every second count like Match or Bridge. They were a little more focused on productivity, like corporations. High Tech High or School 21 meanwhile took as a starting point the outdatedness of schooling, pouring resources into reimagining education for the future. They were flatter and less hierarchical, a little more like cities. Both approaches had real strengths, and I wondered what it would take to combine the best of both in a single model, to systematise but also to humanise? I'd heard rumours of a network of schools in the Bay Area that was trying to reimagine teaching and learning in exactly this fashion, achieving outstanding results for kids. I set off up the coast to investigate.

The System Doesn't Make Us, We Make the System

In a Starbucks on Daly City's Serramonte Avenue, two ex-cops were sheltering from near biblical rain. Flash-flood warnings had been in place across the Bay Area for three days and the downpour wasn't letting up. A younger guy with a buzz-cut had just been suspended for a misdemeanour. He was out. New management didn't get it, they were all targets and procedures. As he drowned his sorrows in a blood orange San Pellegrino, his retired partner was advising him to *retrain, get a new job*. He'd heard Ford were opening up, GMC too. But what was an unemployed 40-year-old to do? Growing up, jobs were easy to come by and he'd not had to worry about school. Now a few miles south-east, driverless trucks were racking up thousands of test-miles on the wet Silicon Valley roads. There was a premium on learning, our only inextinguishable resource.

Driving out of the lot, I heard the US Secretary of Education, Betsy DeVos, on the radio arguing passionately that schools weren't for teachers, but for *parents*. They weren't. Schools were for all of us, shaping our shared future. And, above all, they were for *kids*.

On a rocky outcrop overlooking Highway 280 sat a school that took its work preparing the next generation seriously. Summit Shasta

High wasn't much to look at. Two rows of dilapidated temporary cabins filed either side of a tarmac strip: it was a set of barracks in a car park. An adult-learning centre behind them looked like a penitentiary, obscuring from view the field where the school's new campus would one day be built. 'They've got as far as cutting the grass,' said a teacher. Trees ranged around and the Pacific Ocean could be seen through the occasional gaps in the low grey cloud. It was the America of Ernest Cline's novel *Ready Player One*, a steampunk nation of gamers, side-hustles and urban decay best viewed through drizzle.[16] Yet the scene was alive. Each of the barracks had a brightly painted door. A crowd of Californian teenagers were picking their way between the puddles in groups, skate-shoes sodden, long hair and hoodies wet with rain, still radiating health.

In Mr Davey's eleventh-grade science class learning was already under way. One wall of the cabin was covered with an enormous graph showing atmospheric levels of carbon dioxide since 1959, which students had assembled the previous week as a first act of resistance against a climate-denying government. Nicolas explained that environmental issues were a big concern. The previous project on a water crisis had meant experimenting with water purification techniques. Now 'What is biodiesel?' was written up in blue marker, along with 'How can my team make the highest-energy biodiesel?' and 'Why is biodiesel important?' Thirty teenagers tapped away silently at their computers, striving to solve a problem that would make their world a better place. It was these laptops and this focus that had put Summit on the map. Its story had started two decades earlier when an extraordinary teacher and California native started worrying about our kids.

Diane Tavenner started her career as a teacher confused. 'It seemed so clear to me that there were so many things structurally wrong with what we were doing. We were setting people up to fail,' she told me. When she asked *why*, everyone told her it was how things had always been done. 'Literally,' she added. But she kept on asking. The system seemed resistant to learning. But Tavenner had a different attitude. 'I came to it with a very childlike mind,' she said. That perspective had carried her through her life and career. Today she was CEO and founder of Summit Public Schools and a learning revolutionary. Blonde-haired, piercingly smart and an obvious Californian native, she loved long walks with her family and black Labrador and occasionally found time to read books on social justice, like *Hillbilly Elegy* and *The Warmth of Other Suns*. She'd worked her way up as a teacher in the LA public school system before going on to Stanford, and had spent a further decade studying the very latest learning science. She traced her inquiring mind back to her own childhood.

Back then, Tavenner hadn't been certain if she'd make it to adulthood. If she did, she knew for sure she wouldn't become a teacher. 'I didn't think I would live very long,' she confided. 'That's a fairly common belief when kids grow up in poverty and traumatic situations.' Her father was an abusive alcoholic and her home in the small rural town of Lake Tahoe – where her mum taught an informal preschool – was a place of fear. If she made it out, she only knew that she didn't want to be like them. The village school was her only sanctuary, 'probably the safest place I knew'. It happened to be experimental, open air and progressive, with no classrooms and just a single school hall. She remembered that the teachers eventually built walls around their own areas, which taught her something about the difficulty of making change. Moving to a local high school, she sensed the future struggles of her peers.

'A very small group of kids were on track to college,' she said. 'The rest were not.' It was a calcified, streamed, industrial system that failed most children.

She never stopped inquiring, until one day she had a realisation. The system wasn't designed around the best way to learn, but the most effective way to run the system. It was a revelation. 'One of the key features of the industrial model is that *time* is the fixed variable,' she told me. Traditionally, every child moved at the same pace through the same content to the same rules, in homage to Adolphe Quetelet's average man. It made perfect sense if your primary goal was to 'manage and control an environment', but to Tavenner that was clearly the wrong goal. Instead *learning* had to be the fixed variable. 'All kids,' explained Tavenner, 'given a variable amount of time, and the approach that suits their needs, can achieve the same level of mastery.'[17] It was as Ben Bloom's 2 Sigma Study had shown: with one-to-one support, *all* kids could excel. All the evidence – from Steven Pinker to James Heckman to Todd Rose – said it was worth ensuring all of them did: the trick was how to systematically deliver individually tailored learning at scale.

The problem would have stretched Einstein. With 30 kids in a classroom and no extra money, how did you vary *time*?

Part of the secret lay in those laptops. Starting at Summit in sixth or ninth grade, new students were handed access to everything they would cover in their whole school career, through something called the PLP, or Personalized Learning Platform. In biology class, Nicolas showed me how it worked. Opening a plain-looking dashboard on his Chromebook, I saw tabs labelled 'Current', 'This Year', 'Learning Continuum' and 'Grades' along the left-hand side of the screen. He clicked 'Current' and the subjects he was studying this semester – AP Calculus, AP English Language, Spanish 3, Chemistry – appeared on the screen. For each one there was a headline assignment with hand-in deadline, tabs for practice assessments to complete alone and 'Power Focus Area' tabs shaded green or orange. These summarised how far he'd got in covering the course content. Selecting 'This Year' revealed a timeline, with a set of large blue boxes to represent projects. Below them a couple of rows of smaller green boxes indicated the self-guided assignments. He could engage with any of the modules at any time, make his way through the content himself, then complete the online test when he felt ready.

An ominous blue vertical line reminded Nicolas where he ought to be in his work by now, bringing to mind a line of Meryvn Peake's, 'this desperate edge of now'. He was way ahead.

Could it mean the end of human-to-human learning? I pictured kids at home alone, eyes downcast to glowing screens, making a slow, personalised journey through the content they wanted to learn. Not as Tavenner saw it. To be distracted by technology was to miss the point of Summit. 'It's an essential element. It's not the answer,' she said. Instead, it was all about human qualities. 'One of the things our schools do by design is empower and create relationships, which is woefully absent from industrial-model schools.' There was more. Summit divided the year into four eight-week semesters. Six of these were spent on academic projects, PLP time, and extra maths and language classes. But for the other two, kids had Expeditions, where they could opt into yoga, theatre, art, film, computer science, web design, trips and internships. Summit had looked hard at what it would mean to succeed in college and beyond. Kids would excel academically – 96 per cent of graduates went on to four-year colleges – but they'd also have a chance to find out what they loved to do. Nicolas liked coding and played on the basketball team. He was also keen to talk up the close connections he had to others in his school. I made my way over to twelfth-grade English to learn more.

Inside the portable cabin, faced with 90 minutes of Shakespeare, I was shaken out of a dim Pavlovian urge to rest my eyes by Ms Watts, whose anarchic tattoos foretold a no-nonsense attitude. The class was studying *Othello* and the teenagers were given an hour to work together on one of six group tasks she'd prepared. I took a seat by the bookcase, containing the usual *1984*, *Things Fall Apart*, *The Sound and the Fury*, *Their Eyes were Watching God*. Next to me sat a group of five Asian-American kids in sportswear, led by Jen and Pam. They'd chosen to write a modern-day version of Act 1, Scene 3, although it wasn't clear that they'd all read the play. 'I'm gonna do it in a document,' announced Jen, flipping open a Chromebook thick with stickers from code club. *Hack to the Future! Empire Strikes Hack! Jurassic Hack!*

Here in the Valley you learned to code like you learned to read. The others followed suit.

'Different colours, different characters,' added Pam. The blank Googledoc began to fill with sentences in five different font colours as they typed away at the same time. They were good at this.

The scene saw Othello, after seducing Desdemona, scorned by Brabantio. The five of them decided basketball and NBA owners would be their setting and they quickly divvied up the characters.

'How about Des? Or Destiny?' asked Alex, looking to rename Desdemona. The basketball players Steph Curry and LeBron James were written in, along with the R&B singer Usher.

I noticed that one of them had written, 'I slid into her DMs!' What did that mean?

Pam laughed, setting off the others. They promised me it wasn't rude, but meant making a move via social media.

'Where did Usher come from?' asked Gabi.

'Slid into her DMs! *Icantbelieveyouwrotethat!*'

The scriptwriters collapsed in hysterics, before regaining their composure and typing on, pausing occasionally to cackle at or adjust a line. This was new to me. It wasn't that the output was particularly remarkable (though it was funnier and more original than anything I'd written aged 16), what stood out was their absolute comfort with the process. It was hard to evaluate creativity or co-operation, but more than any other kids I'd met, I felt like these five had it. They were relaxed, generating ideas, thinking divergently, then doing something to refine and craft what they'd come up with. And they were doing it together, multiplying each other's abilities. This was thinking anew, doing better *and* taking care. In 15 minutes, they filled two pages, bringing the script together. As the deadline approached, they were done. Was this how writing teams worked down the road in Hollywood?

'Send me the link to that,' asked Ms Watts. I was a little in awe. In the course of an hour, the five twelfth-graders had taken a project brief, assigned team roles, reviewed a brief text and then co-written a punchy, imaginative and *funny* two-page script. I thought of the endless hours of office meetings that went nowhere, the drafts and redrafts emailed round to managers and senior executives. These kids had some real-world skills. They'd have a great shot at the jobs of the future. They'd also have a great shot at life if the robots came and we no longer had to work. The five kids were well equipped to rise to a challenge, make a situation work for them. Best of all, they found joy and camaraderie in the process.

'If she hath indeed allowed her DMs to be slid, suspend me from my own game!' The writers creased up with laughter.

On the wall, a C. S. Lewis quote said: 'We read to know we are not alone.' I understood. Socrates had seen something in reading that we now saw in our smartphones. You accessed a new virtual world but disengaged from people. We had to balance the poison and the cure. At Summit kids spent 16 hours a week in PLP time, eyes on a backlit screen, reading, watching, writing, taking content tests at their own pace. Their textbooks were open source, kept up to the minute by their teachers, who constantly tweaked, updated and added new projects. That was part of the system's beauty, never becoming outdated. Beyond that they were in project teams, thinking together, helping one another out. They were practising independence, learning responsibility. 'Words are our most inexhaustible source of magic,' said another sign, quoting Harry Potter on the power of poetry to remind us of our irreducible humanity. 'They are potent forms of enchantment, rich with the power to hurt or heal.'[18]

Tavenner stressed this humanity as we spoke. Summit had been running since 2003, when she had set up the first school in Redwood City near Palo Alto. Though it had been an immediate success, in 2011 she had made a discovery similar to Dave Levin's at KIPP. Though 96 per cent of her kids were going to college, only 55 per cent were on track to graduate on time. It was a punch in the gut, demanding a complete rethink. 'We had set up this environment that had very high expectations, but also very high support for kids,' she explained. Once they departed they didn't have the same network and lacked the know-how to meet new challenges. 'Skills like that only come with years and years of practice,' said Tavenner. She went back to the drawing board, realising that in a school environment valuing compliance, it was impossible for kids to learn things like independence or judgement. Everything had to change.

'It's not a project you can do in your senior year one time and "Whoop!" you're ready to go,' she said. I'd seen in Shanghai how policymakers were wrestling with the paradox of a high-performing but authoritarian system. In Finland, it came more naturally. 'What is interesting is that once you have that realisation and you commit to it, *everything* changes,' she continued. 'You have to rethink everything you are doing in school, because it is completely oriented around the adults controlling and managing the environment.' Just like in Geoffrey West's analysis of corporations. Summit decided instead to equip kids to succeed on their own. As a result, they made as big a deal about habits of mind as they did developing cognitive skills, adding extra maths and English classes to keep kids' levels high. Technology supported that, saving teacher time, allowing kids to go at their own pace.

Tavenner had only hired Summit's first developer in 2013. He'd got no further than 'duct-taping' a system together for them to use when Priscilla Chan came to visit, and, later, Mark Zuckerberg. The tech-gods smiled on what they saw and right away Zuckerberg gave Summit's developer, Sam Strasser, a team of Facebook software engineers to help refine his system. The PLP was born, soon followed by Basecamp, a platform Summit had developed to share the PLP with the world, free of charge and unrestricted. Tavenner called it 'the operating system of the schools of the future'. If Mr and Mrs Facebook were backing it, she probably had a good chance of being right. Basecamp was already used by hundreds of schools across the US.

Looking at the stained and missing ceiling tiles, the basic Chromebooks, the puddles of water forming on the strip of tarmac, it was clear that this model would be replicable almost anywhere. You needed teachers, peers and laptops. It was the teachers that were hardest to come by. Summit's system was designed to grow them too. The capacity saved through Personal Learning Time and Expeditions was used to put teachers through 50 days of professional development and peer collaboration every single year, unheard of in most places. Summit was building a supportive infrastructure around *human* systems. 'I actually have a huge belief in teachers and school leaders,' said Tavenner. You just had to free people from serving the system, save them from having to swim upstream or get beaten down. You had to help teachers become expert learners, fuel professional communities, harness the multiplier effect of cities. Technology was just a tool. Learning lay in *us*. We were all natural born learners – teachers too.

In its early years, Summit had regularly struggled for money. At the lowest point Tavenner summoned a meeting of all the kids and their families to attend a public hearing to prevent the school's closure. She told them she had failed them, that she hadn't delivered on her promise. Her voice trembled as she recalled the day. 'This one student stood up and said, "You know what, don't worry about it, Ms Tavenner. This school's not based on place. It's based on relationships. And I don't care if it has to be in a park, or a parking lot, we will follow you. This is our community. We're together."' It was the moment she knew that Summit had succeeded. 'I realised then,' she said, 'and I think it's still true, that the schools we've created are so *impersonal*. They're not places where kids know that anyone loves them, or cares about them. And they're not stupid. They know that we're not getting them ready for the world they are about to enter. And the fact that we *were* able to do that negated the fact that they literally might have to learn in a parking garage.'

Kids knew that other people mattered most, cared that others cared. Learning grew from the multiplier effect of human connection. Just like in cities. Or minds, for that matter.

Rethinking Our Learning Machines

West's research into cities reflected new advances in AI. What if a machine could learn not according to a set of rules that it was forced to follow, asked Alan Turing in the 1940s, but from sensory evidence and scientific inquiry, just like our own brains?[19] For 70 years boffins had ignored his idea, programming computers precisely *to* follow rules. *If A, then B. If X, then Y. If this, then that.* Each element of a computing system was assigned a function and told to repeat it over and over again. This rule-following could be deeply impressive, enough to win games of chess, beat Ken Jennings at *Jeopardy!* or examine X-rays more reliably than radiographers, but it was still essentially a sham. The machines, as Gary Kasparov had pointed out after losing to Deep Blue, were no more intelligent than our digital alarm clocks. They mimicked aspects of brainpower, but were incapable of complex thought. Experts called this 'weak', or 'top-down', AI. The 'bottom-up' sort was presumed to be impossible to create.

Then in November 2016 a Japanese professor of computer–human interaction called Jun Rekimoto noticed that the quality of Google Translate had leapt suddenly overnight from tourist-reading-from-phrasebook to poetry-in-translation. He decided to test it, feeding it a passage from a Japanese-language translation of Ernest Hemingway's *Snows of Kilimanjaro*. Well accustomed to the clunky wording of old Translate, Rekimoto awaited the usual obtuse rendering. Instead, it proposed, 'Kilimanjaro is a mountain of 19,710 feet covered with snow and is said to be the highest mountain in Africa. The summit of the west is called "Ngaje Ngai" in Masai, the house of God. Near the top of the west there is a dry and frozen dead body of leopard. No one has ever explained what leopard wanted at that altitude.' It was near-literary.[20] Pushed to the margins since Turing first posed his question, 'strong' artificial intelligence had finally made its breakthrough.

Translate's new powers lay in an idea that had first occurred to the Cambridge code-breaker himself, that of the neural network.

Designed in the image of the human brain, the neural net takes *natural* systems as its inspiration. In a brilliant piece for the *New York Times* which recounts this story, Gideon Lewis-Kraus compares learning by the neural net to evolution, in contrast to the workings of weak or top-down AI, which resemble creation. He writes that just as within our own brains 'with life experience, depending on a particular person's trials and errors, the synaptic connections among pairs of neurons get stronger or weaker', so an artificial neural network 'could do something similar, by gradually altering, on a guided trial-and-error basis, the numerical relationships among artificial neurons'.[21] If classic artificial intelligence worked a bit like a corporation, then a neural net functioned a lot like a city.

I saw in it an analogy for how we might unleash our own unlimited learning potential. Couldn't we create strong systems of learning around a bottom-up model?

Before setting off on the last leg of my journey, I'd spent time talking to global education leader Wendy Kopp. Six years earlier, I'd left the classroom to join Teach for All, a global network of organisations established a decade ago to mobilise the leadership of teachers, kids, parents, principals and policymakers to improve learning in their communities for everyone. Kopp was the CEO, and my boss. At college, she grew concerned about the inequities of opportunity in her country and sensed around her a 'generation of people who I thought would love to be a part of making a difference', she told me. 'And what better way to make a difference than through teaching in underserved communities?' After graduation, she had launched Teach for America, a non-profit whose mission was to enlist, develop and mobilise the country's future leaders in a movement for equity in education.

For three decades, Kopp had worked tirelessly to transform learning for the poorest kids, first in the US and more recently worldwide.

After aspiring first to 'close the achievement gap', just like Mike Goldstein, she had come to embrace the idea of collective leadership, working alongside communities that would come together to reimagine education for their kids in the twenty-first century. Across her career, she'd seen governments swing from one solution to another and then back again, from charter schools to voucher programmes, teacher quality initiatives to curriculum reform. 'You could name every year for the thing we were talking about, the silver bullet we were chasing,' she told me. 'Now we're back to project-based learning.' Kopp had long ago concluded that there was no silver bullet, no single superhero about to fly in to lead the recovery of the system.

This was the limitation of the 'top-down' model. It wasn't that Starbucks coffee was bad (I drank a lot of it while writing this book), but that it supposed we required a Howard Schultz to call the shots, that the system would function best if each person just did as they were told by single leader. It was an outdated idea forged in a different age. Top academic and policy expert Anne-Marie Slaughter recently wrote that 'the idea of one "leader of the free world" will soon come to seem very quaint indeed'. Instead, she argued that we should embrace 'a different form of leadership', resembling the neural net, 'of empowering groups to take and implement collective responsibility for tackling specific problems'.[22] The system wasn't out there, some abstract entity. It was *we* who made the system. Human development therefore depended on increasing the capacity of *every one of us* to think anew, to do better and to take care. If we relied on a great leader, we'd never realise that potential.

'We should be working together to reinvent the system,' Wendy explained, 'all of us.' This meant unleashing the collective power of people to decide what would work best in their context. It began with convening diverse grous of local people from every walk of life and asking a simple question: 'What do you want for your kids?' Wendy said, 'Our system is built around people asking that question two hundred years ago.'

As top-down artificial intelligence was a pale imitation of bottom-up, so the same was true for our systems of learning. It was no good for a single leader, or a group of policymakers or academics, to issue a set of instructions for everyone to follow. In an endeavour as complex as human development, there was little hope the approach would be right, and, robbed of their agency, people in the system would do no more than comply, losing motivation and the chance to learn and grow. Instead, it was clear that we'd do better by engaging everyone – kids, parents, teachers, policymakers – in a shared effort to improve learning for all, testing approaches to achieving the outcomes we all wanted for our kids, feeding the evidence into our decisions about what to teach and how to teach it. Our systems of learning had to become self-learning systems.

'*That* can bring us together,' continued Wendy. 'We need to hold ourselves to working together in pursuit of something much bigger than any of us.'

We were born to learn together. We had to rebuild our systems around that insight, from the bottom up, in the image of cities and the neural net, where the power of the whole came from realising the full co-operative potential of each individual element. True learning systems were open source.

The Future of Learning is Missing

Back at High Tech High I wandered in through the fire escape door of a long nondescript hangar, one of the windowless ex-military facilities the school occupied. A row of two-storey-high glass-walled classrooms filled half of the space, opposite a large black curtain that ran the length of the building, as though we were backstage. Attached to the wall was a huge mechanical sculpture of 30 interlocking bicycle wheels, which could be turned by a crank. A tall cabinet displayed detailed wooden models of Escher's tessellations and unending staircases. In an enclave, students were assembling two-storey staircases and geodesic domes in pine. Sunshine flooded in from the skylights above, through the life-size portraits dangling on cord from the ceiling. The scent of summer and sawdust filled the air. *The Garden of Earthly Delight* read a delicately painted sign.

In the first atrium, Zack and Ava were manically working the remote control of a home-made drone, one, two or four of the motors jerked into life after a short delay and whirred apologetically. The miracle of flight was still some way off. A rivet popped off, sending a propeller skittering to the floor. Energy drinks and a half-eaten raisin bagel sticking out of a tub of cream cheese suggested it had been a long day. Zack scratched his mad scientist curls. 'It has to be operational by Friday,' he said. He was learning something about drones, and something else about deadlines. Next door, another group was further along preparing for a test flight. This semester these twelfth-graders spent half their time on this enviro-science project and the other half in humanities and calculus preparing for their SATs, the US college entrance exams.

Amid the chaos of the two drone-building teams Zoe sat with her headphones in, typing. She hoped to go to a liberal arts college in New York to study theatre, and her role in the project was to write the treatment for a making-of documentary. Two further documentarians, Emir and Jim, were editing footage they had shot earlier in the day on their laptops. Outside, the drone-pilot Amachai was practising. A group in lab coats were studying seeds. They'd prepared a series of soil trays and were testing different seedpods built from clay, earth and biodegradable plastics. These would be scattered by the drones, Jim explained. Once prepared the project team would fly drone-mounted cameras over Cabrillo Monument National Park to survey the biodiversity and identify any early signs of desertification. California had a lot of droughts. If any sites seemed lacking in the usual flora and fauna, they'd fly the drones over a second time and drop seeds on those areas. It would require a four-day hike into the wilderness, the whole adventure captured by the documentary crew.

'It's going to be *awesome*,' said Jim, 'we're going to be taking dumps outside in holes.'

My sixth-form essays on *The Waste Land* and *Cyrano de Bergerac* had felt important to me, but this felt rooted in the material world. Down the corridor, a teacher, Jeff Robin, known affectionately as 'the Lunatic', pressed play on his latest Spotify playlist and a heavy guitar riff chugged. The multidisciplinary collaboration was important in an era demanding co-operation, but lacking interconnection between areas of expertise. High Tech High's 16- and 17-year-olds spent half their school day experimenting with environmental activism, which ultimately required planet-wide co-ordination. They weren't painting imaginary placards or writing essays though, but doing it for real. A laser-cut wood panel on the wall summarised the ethic:

A thinker sees his own actions as experiments and questions, as attempts to find out something. Success and failure are for him: answers above all.

Friedrich Nietzsche

It wasn't about computers here either. 'The name's a bit of a misnomer,' agreed Rosenstock back in his office. A National Geographic Explorer of the Year, newly returned from a search for Genghis Khan's tomb, was just making his way out. 'Nothing's new,' continued Rosenstock. 'The technology is just a tool.' The name suggested it was all about iPads, but touring the school it became clear teachers mattered most. 'They are all in teams,' he said. 'Teams, teams, teams.' Not for them the closed-door shame or 'autonomous isolation', but teamwork, close-knit community, open doors and 'tons of professional development'. Rosenstock thought carefully about ensuring staff were successful. They met in the mornings, rather than afternoons. 'People who meet after school talk about what Billy just did wrong. People who meet in the morning talk about next year I want to do this.' The school's reputation meant they now had their pick of the teaching talent: 1,800 online applications were made each year for their vacant positions. 'A clusterfuck,' he added.

High Tech High had its own teacher-training programme. Along with an old colleague from Harvard, Rob Riordan, Rosenstock had noticed that of the 1,400 graduate schools of education in the US – where all the country's teachers were trained – '*none* of them are in schools'. He was incredulous. 'Not having graduate schools of education in K-12 schools is like going to medical school and not seeing a body.' So they'd built their own. Again, it was learning by doing. Teachers chose an inquiry question, plotted classroom action based on the latest theory and then went for it. It was a craft. They believed in socially constructed knowledge – Freire and Socrates – but also in deliberate practice. I'd bumped into one of the trainers outside, who told me that the Mr Miyagi-ish jazz master who'd taught him piano told him that five minutes' *perfect* practice each day was better than hours of half practising, half paying attention.

The culture of learning in the school was spreading out into the world. It was proudly open source.

Was it the school of the future? Rosenstock answered my question with a parable. At the end of the New Urban High School project, he'd presented at a meeting of the governors of every US state. Instead of describing the ideal school, he'd told them about a film. *Chan is Missing* was an Eighties art house movie made on a shoestring budget in San Francisco. It told the story of a pair of down-at-heel taxi drivers in search of an old man named Chan to whom they'd given money to get cab driver licences. Over the course of a weekend, the two protagonists undertake a quest through the city's Chinatown, encountering a cast of local characters in the evocatively imagined baths, tea-shops and restaurants of the neighbourhood, each of whom gives a personal and often contradictory account of Chan and his motives. 'Who is Chan?' we wonder, all the way to the end. The film closes with the old man's daughter returning the money to the cab drivers and handing them a photo, Chan's face obscured from the audience by a thumb holding the picture.

I thought about my journey. High Tech High and Summit weren't *the* school of the future, any more than Hiidenkiven Peruskoulu or Wanghangdu Road Primary, 42 or King Solomon Academy. Larry Rosenstock and Diane Tavenner didn't have the single answer any more than Daisy Christodoulou or Kathy Hirsh-Pasek, Andreas Schleicher or Joshua Wong. It wasn't yet possible to lay eyes on the perfect school or system, just as you could never see Chan. Instead, the inquiry mattered. You began to see who Chan was through the way people interpreted him. At each turn of my journey I'd found people and places striving to ensure kids reached the limits of their potential. What distinguished them from others was learning itself. Each of them had undertaken to think, try, reflect and learn continually about what it would take for kids to get there. More than their philosophy, what mattered was their commitment to excellence. As Wendy said, the more everyone – kids, parents, teachers, principals – played a shared role in that learning, the more we'd multiply our impact.

I'd started off on my journey hoping to discover how to change schools. Now I realised we had to start with ourselves.

We were natural born learners. If we outsourced thinking to the machines, we diminished the power of our brains. If we built institutions of learning around authority and hierarchy, we lost the potential ingenuity and capacity for continual growth of everyone in the system. Learning was about knowing, doing and being. The paradox at its heart said you had to be told how, or you would never think, never find shared ground and never achieve expertise. It also said that you couldn't *only* be told. Who then would have the spirit to advance our wisdom, push forward our culture, look far into the future to solve our biggest problems? You had to experiment, to trial and above all to fail. We thought *together*. To learn was to be human. To err was too. The computer age wanted to eliminate risk from our lives, aspired for everyone to be the same. But we needed risk. Risk meant failure. We had to fail together, fail better. It was the only way to learn. And there was only one place in the world where we could do that safely.

I made my way out through the gates of High Tech High and towards the vast blue ocean glinting in the early-evening sun.

In my search for the future of learning, I'd looked to artificial intelligence, neuroscience, early-childhood development, creativity, character, lifelong learning, teaching and democracy for some glimpse into what was to come, a metaphor to capture the importance of learning. Gazing out at the darkening Pacific, I realised I'd been holding my thumb over the obvious answer. *School itself* was humanity's greatest and most important invention, the place where we came to cultivate our species' most precious inheritance. *It* was the means through which we advanced our culture and our technology, the fuel that powered our societies. We had to build on our successes and reimagine its purpose for the next generation by thinking anew, doing better and taking care. In ancient Athens it was for noblemen. In Shakespeare's time, the sons of glovers. Today every child – a billion of them – had access to it, though many of them weren't learning. But we *were* getting closer. We now knew the ingredients of great learning. The stars were brightening in the night sky. We had to build around people, not technology. Embrace solidarity, not competition. Develop our performance, as well as our intellect and ethics. Politics as much as economics. In the age of the internet, *this* – learning – really was for everyone.

AFTERWORD

A Learning Revolution

> If you're not learning,
> you're wasting your time.
>
> Pharrell Williams[1]

Education is the Silver Bullet

I've lost count of the number of times in the past decade I've been told that education is the answer. What can we do to tackle poverty? Education. How can we bring about gender equality? Education. The solution to our population explosion? Education. Robots coming for our jobs? Education. Wars displacing refugees? Climate change threatening the planet? Society coming apart at the seams? Education, education, education. Learning is our everything cure, the secret to health, happiness and global co-operation.

If you're waiting for the 'but', it's not coming. Education *is* the answer. Schools aren't to blame for the ills of our societies – we must redistribute wealth, tackle poverty at its roots, create a fairer, more ethical and sustainable world, alter the ideas that govern our actions – but they are the solution. We need creativity, compassion and co-operation to tackle those challenges. That means realising more of our human potential, ensuring future citizens are better equipped than us to build a better world for everyone. Learning is the cause of our generation. As Aaron Sorkin wrote in *The West Wing*:

Education is the silver bullet. Education is everything. We
don't need little changes; we need gigantic, monumental
changes. Schools should be palaces. The competition for the
best teachers should be fierce. They should be making six-
figure salaries. Schools should be incredibly expensive for
government and absolutely free of charge to its citizens, just
like national defense.[2]

He's right. Every indicator of well-being or wealth says that education *is* our best investment. Finland, South Korea and Singapore have
demonstrated the returns. In half a human lifetime each underwent
a school-powered social transformation to become world-leading
technologists, thinkers and global powers. History tells a similar
story. The sixteenth-century invention of grammar schools gave us
Shakespeare and the English Renaissance. Plato's Academy gave the
West an enduring philosophy. We learn therefore we progress.

I'm convinced that we're on the cusp of a further renaissance.
We need one. Of over a billion kids in school around the world, 600
million are falling behind.[3] Our systems resist change, entrenching
inequality, poverty and undermining well-being. Updating them will
be an epic endeavour. Yet in a globalised world where everything is
running short, a learning revolution is our one shot at flourishing
as a species. As land, food and fuel dwindle, our human powers
of thinking, doing and caring are our sole unlimited resource. We
must cultivate them in their fullest extent. 'If your plan is for one
year plant rice,' said Guan Zhong. 'If your plan is for ten years plant
trees. If your plan is for a hundred years *educate children*.'[4] We know
now that we are all born to learn.

But what precisely *should* we learn today? What will it take to
realise our potential, to recognise school as our greatest invention,
to embrace teaching as our ultimate craft? How should we define
success for future generations? What will it look like to rebuild our
systems around the way we learn, rather than how we run schools?[5]
Technological and scientific revolutions have transformed our world
since Plato's day. Our journey has shown that we can use our latest
advances to rethink what we value, retool our education systems
and fire a learning revolution. We must begin by reframing ideas that

tie us to the past, rewriting the stories we tell. The revolution starts here, with a manifesto.

1. Learn Forever

It begins with learning. Education once entailed the transfer of the sum of human knowledge from one generation to the next, but today our institutions develop too slowly to keep pace with the rate at which knowledge grows. It no longer makes sense to front-load learning at the start of a life. We old-timers no longer have all the answers to pass on anyway. At 42, Nicolas Sadirac, who is one of France's top tech experts, admitted students at his school could now do things with computers that he couldn't comprehend. Instead of teaching our kids to know, we need to teach them how to *learn*.

We are born to do so. The genius of our human minds is that they are endlessly adaptable and more powerful than we realise. Learning is our superpower, but rather than unleashing our innate potential, our models of education are too often limiting it, framing our minds as computers to be fed information, reducing learning to a program of inputs and outputs. Human intelligence is not fixed, however, but alive and in perpetual change. We must learn to wield this evolutionary inheritance throughout our lives, growing our cognitive powers.

Self-motivation is the key, as Ifrah Khan and Lilas Merbouche showed, but we too often kill it. Every child is innately curious and uniquely individual. As Pekka Peura said, we have to delete the idea that learning equates to passing standardised exams, and rebuild our approaches around fuelling kids' ability to wonder, to imagine, to express ourselves, to analyse, critique and question, to inquire like scientists or learn simply for the love of the learning itself.

'Training someone early to do one thing all their lives is not the answer,' Andreas Schleicher told me. Future generations face long, uncertain lives of multiple occupations. If they're equipped to be lifelong learners, then they can do so with confidence. We can ensure they understand how to explore options, find their sense of purpose, practise growing expertise in a chosen field and learn to work in teams with others. As George Saunders urges, 'Stay so open it hurts.'[6]

2. Think Critically

That doesn't mean giving up on knowledge. Shakespearean English, Newtonian physics, Euclidean geometry and historical fact remain the basis of our shared Western culture, the anchors around which we build our cognitive abilities and root ourselves in the human world. But the purpose of learning this information is not to win *Jeopardy!* or *University Challenge*, but to lay the foundations for thought.

The ability to think critically matters more than ever. Individuals and organisations are out to hijack our minds, subtly nudging us to achieve someone else's aims. YouTube and Instagram harvest our attention. Advertisers seek to shape our behaviours. Misinformation erodes our shared culture. How then do we learn, as David Foster Wallace put it, 'to exercise some control over how and what [we] think'?

There are no shortcuts. Ifrah's experience at KSA shows us that continual struggle is a core requirement of this process. If it feels easy, then we're avoiding real thought. That can be our test. In classrooms, our kids ought to adopt the inquiring, doubting stance of the scientist. As Benny Tai told me, teachers can grow this capability in their students by encouraging their students to question what they're telling them, to question their own beliefs, and to challenge those of their peers.

Mr Hartley played this role in Philosophy for Children at Gallions Primary School, gently pushing the little learners to examine the roots of their opinions, and prompting them to create shared reference points as a group. Public discussion and debate are our path to resisting our own biases, and the manipulation of our minds by others. Schools are the antidote to the echo-chambers found online. Let's bring up our kids so that they can't be cheated.

3. Get Creative

Kids are born free, but everywhere are in schools. At Walworth I was so focused on mark schemes and behaviour plans that I never stopped to reflect on the part I could play in fuelling imagination, sparking interest, or fostering creative abilities among my students. My own narrow academic success blinded me to the importance of *doing* better, or breaking the rules. Let's forget about compliance and instead teach creativity.

That puts craft back at the centre of human learning. As in Hiidenkiven Peruskoulu, School 21 or the MIT Media Lab, doing so means making more space for our kids to play, giving them freedom to experiment and fail and offering them many chances to find their passion. Most of us went to schools where we made art, performed in plays, carried out science experiments, indulged our curiosity, but too often today, our kids are missing out on these experiences.

We can ill-afford this lost potential. Not only are the creative industries one of the fastest-growing economic sectors, but as our population grows, resources become increasingly scarce and our ideas run dry, creativity is also vital in ensuring we can think anew – and think big – as a species, taking on our greatest challenges of over-consumption, climate change and environmental degradation, and doing better in tackling them.

That creativity can be developed in any media, whether writing, art, music, drama, maths, science, coding or sport. It's a fundamental, not a nice-to-have. The tools kids develop in these pursuits will help them find purpose, and give them lifetime access to the uniquely human trait of self-expression. If mastery of core subjects means we're not finding time for the arts, imagination and creativity within school, then we must find places for them elsewhere. A child's spark might change the world, could be their career, or may simply become the place where they find meaning or joy. This is OK.

4. Develop Character

Building a healthy society means ensuring our kids are well nourished in both body and mind. Yet today our young are in the grip of a mental health crisis. One in four girls in the UK is clinically depressed by the age of 14.[7] Worldwide, 450 million people suffer some form of mental disorder.[8] Education is a deeply human endeavour. So along with the intellectual and practical, let's take the emotional development of our kids seriously. What message are we sending to them if we don't prioritise well-being? What does this say about the type of society we choose to live in?

They can learn to be resilient. Angela Duckworth's research and the examples of KSA, KIPP or Korea prove that kids can be equipped to succeed against the odds, carrying that success on into their lives. This approach requires tough love, the highest expectations and the creation of a no-excuses culture. It also requires trusting, loving relationships to be established among members of a community. Kids that are more secure in their attachment to others are better able to cope with life's challenges.

Feeling part of a community is fundamental to kids' well-being and mental health. To grow a healthier society, we need to help kids unpick the forces that are shaping who they are and how they behave. Boys and girls alike can become more emotionally literate, recognise parts of their gender as constructs, free themselves from damaging expectations they've internalised about beauty, bravery, intelligence, leadership, humour or love. Schools should be a safe space for risking new opinions, trying different identities, learning to love.

Mindfulness offers us one path to achieving these outcomes. At Breakthrough Magnet I saw the power of putting well-being at the centre of the education mission. Not only were kids learning to think and to do, but also to feel. In mindfulness classes students and teachers learned to listen to their emotions. Through tested psychological techniques, they learned to build a happy, loving community. We can make it the work of all schools.

Let's put mental health and lifelong well-being at the heart of our learning revolution. Even as we prepare our kids to handle the race to the top, we can do still more to help young people find contentment and make sure self-esteem isn't sacrificed for mindless competition. At KIPP Infinity, Jeff Li was not afraid to tell me – or his students – that he loved them. This should be the norm.

5. Start Early

Throughout my life as a preschooler, pupil and student, it was clear that the older I got, the more important learning became. I was dead wrong. Maternal-care nurses at the Perry Preschool Program in

1960s Michigan added more in one hour to the life chances of the tots they visited than their elementary and high school teachers later managed in weeks of classes. This insight must define our systems of learning.

Our old hierarchy says university professors trump college lecturers, who in turn lord it over school teachers. Sixth form then beats secondary, which is better than primary. At the bottom of the pile, after dinner ladies and facilities managers, comes the motherly cadre of nursery practitioners. This is bad thinking, ranking roles by the complexity of the knowledge the learner is acquiring rather than the complexity of their development. It is defunct and should be flipped.

In Pen Green, I witnessed a new way. Expectant mothers attended ante-natal classes, continuing to drop into the centre once their kids were born. From a year, babies could begin spending half-days at nursery, where they'd benefit from specialist care from highly trained staff, tailored to their developmental needs. Attached to Pen Green was a world-leading research centre. And they hadn't shipped in experts from different corners of the globe. Most of the women completing degrees and PhDs were local mothers from Corby, one of the UK's poorest and least educated towns.

Let's imagine a system that ensures the environment around the youngest learners is designed to maximise their lifelong success, committed especially to supporting mothers and fathers most suffering from deprivation. We currently make only the smallest allowance for these services, yet money spent at this stage of a child's life has the biggest possible return throughout their lifetime.

6. Grow Co-operation

Young people today are more socially conscious, altruistic and clean-living than their forebears. Generation Me is giving way to Generation Us. Yet from aged five – and often even younger – we pitch kids into a brutal race to the top, tell them to conform to a certain standard, and then rank, sort and select them. This made sense in eighteenth-century Prussia. It makes sense in the business world today. But things have changed.

Hierarchy is no longer the best way of doing things. The world is too complex. Situations change too fast. Instead, we should be teaching our kids teamwork, building community. In these human systems every person plays a key role in the success of the group, so it becomes vital that we maximise the potential of every individual, and learn to collaborate. Instead of figuring out 'Who is the best?', we must ask 'What might we be able to offer each other?' and 'How best can we work together?'

Learning built around an idea of team and community is inclusive and democratic. Rather than pitching kids against each another to achieve a one-size-fits-all criterion for success, we should personalise approaches to individuals, allowing them to proceed at their own pace and ensuring that everyone learns. Pekka Peura did this in Martinlaakso. Todd Rose advocates it in *The End of Average*. This means allocating more resources to anyone who struggles, or has particular educational needs. We should do this gladly.

It can't be an excuse to lower expectations. The race to the top has lifted levels of learning, created new models of high performance. We must use the best of these to close gaps in education levels, rather than open them. More-equal societies are also healthier, happier and performing better on almost every metric. That doesn't mean everyone within them is the same.[9]

7. Practise Teaching

At a recent talk Georges Haddad, the director of France's most prestigious university, La Sorbonne, said something surprising. A physicist by training and a grey-haired French academic, he launched into an impassioned speech about young people. 'A society that doesn't love its teachers,' he concluded, 'doesn't love its children.' What kind of society would ever say it didn't care about its kids?

In 50 years, we'll treat teachers like doctors. They'll train like athletes, research like scientists. They will be masters of learning, the role to which everyone aspires. In a future defined by the jobs that humans do for one another, it will be our ultimate craft. This is already becoming true in the world's best schools and systems. In Finland and Singapore teachers are proud, autonomous and well-trained professionals, masters of a learning profession. It was true in School 21 and High Tech High, Summit and King Solomon Academy.

We're some way from achieving this goal. Globally classroom practitioners are turning their backs on the profession, overworked and underappreciated. Almost 500,000 qualified teachers in the UK have left to find jobs in other sectors. To achieve global development goals, we need 69 million new teachers worldwide by 2030,[10] doubling their number from current levels. There'll be close to 1.5 billion kids in school. If Osborne and Frey, the Oxford economists, are right about which jobs will be automated, teaching is one of the areas we can create huge value and enriching work.

At Walworth, I saw teaching could be the ultimate craft. Mr Jahans and Ms Toworfe made kids think, unlocked their intrinsic motivation, inspired them to greater heights than they believed possible. So did Pekka Peura in Helsinki, Adie Capaldi in New York and countless other teachers I met on my journey. This was not magic. They had mastered the science of cognitive development, the practice of community-building, the psychology of motivation, the in-depth knowledge of their subject. That expertise can be grown.

To make the most of our teachers, we must trust them, train them well, give them autonomy, free them of the burden of administration and put them in charge. Brent Maddin imagined different roles that teachers might play in the future. Subject experts would teach kids while continuing to research at the cutting edge of their subject. Learning coaches might grow kids' motivation and ability to work effectively with peers. Data analysts might more efficiently find the gaps in kids' learning. We can trust them to rise to the challenge.

8. Use Technology Wisely

The robot teachers aren't coming yet. In Silicon Valley, I'd expected to glimpse a gadget that would radically alter learning for ever. But although tech is changing the world, the routine, repetitive, rule-following types of tasks that computers have mastered are the opposite of what we need in our classrooms. Learning can't be like Angry Birds.

Technology is just that, a tool. We'll best use it to advance our learning when we put it to human purposes. At 42 I saw a school without teachers, where the software was running things. Mitch Resnick at MIT had created Scratch, a teacherless online learning community. At Summit Public Schools kids had a full day every week alone in front of a laptop working through assignments on their Personalized Learning Platform. These innovations created efficiencies, tailoring learning to kids and saving teacher time. They will spread.

Connected computers can also bring learning to communities without schools. Pratham, an Indian NGO working to build the literacy skills of more than 3 million kids in Uttar Pradesh, is experimenting with tablet-based after-school learning groups that are proving effective in developing kids' English. Bridge International Academies, with some controversy, is now educating more than 200,000 kids in East Africa and India using its preloaded teacher and student tablets.

Human plus machine plus better process wins. One of our aims in education must be for our kids to learn to use the tools of today. It means experimenting with tablets, trialling activities on our phones. It also means understanding the tools. As every kid learns to read and write, do maths, think scientifically, so they should learn the basics of programming. We have to beware the dumbing-down effect of outsourcing our thinking to machine minds.

Ultimately most jobs in the future won't require specific tech skills, just as they won't require you to read or multiply beyond a rudimentary level. And if the robots do take the jobs, it's our human qualities that will count. We'll service one another's needs, nutritional, intellectual, spiritual or physical. The greatest impact of technology on learning may paradoxically be to push us towards the human. Let's prepare our kids for it.

9. Build the Future

Finally, let's not wait for the future, but bring kids up to create it. The robots may be about to eat all of the jobs. Hyper-wealthy tech overlords could be set to amass all the riches. Climate could change irrevocably. Global population levels may rise beyond control. We could wait and see what happens, hope for a leader to save us. Or like Joshua Wong, Lilas Merbouche and Da'jia Cornick, we could start learning to work together to imagine and create what we wish to see.

We must bring up the next generation to do so. At Eton College, the UK's most prestigious school and alma mater of 17 British prime ministers, I heard of a particular biology teacher who, as a climate change activist, had deliberately sought to teach at the school. Her reasoning was simple. It was the biggest impact she felt she could have on preserving the planet. Passing through her classroom were future MPs, judges, newspaper editors, CEOs and diplomats. Under her guidance they'd all leave climate activists. If you want to change the world, start with school.

Rather than preparing for a future we can't predict, our kids can create it. What will it take? At a school in Mumbai I met six young students who'd grown up in slums but were now heading off to prestigious universities. They planned to return to their communities to teach and create businesses. School 21 in London asked its pupils not to ready themselves for the rest of their lives, but to positively influence the world around them today. A dozen teenagers at an organisation called The Intersection in Baltimore led a campaign to elect a pro-Dreamer politician.

Today's kids can surf the uncertainty of our age. If we can equip them with the right knowledge, skills and attitudes, they'll figure out together how to build a better future for everyone. Learning isn't the solitary act of developing one's own faculties, but a shared effort of advancing our societies.

We are the System

Though education is the silver bullet, there are no silver bullets to smooth the path of our kids' learning. I have searched the world for the technology or innovation that will transform education. Instead, I found that the future of learning lies in us. This is our cause. 'The most important infrastructure we have is educated minds,' said Amel Karboul, spokesperson for the Education Commission, a UN body established to improve learning worldwide.[11] Strengthening that infrastructure is both the means and the end of our mission.

The 'system' isn't over there. The system is us, made up of the relationships between people. Everyone has a role in it, the power to influence how it functions, for the better, or for the worse. If all of us are innately capable of learning and adaptation, then things as they appear now are not how they must always be. We can and must make a shared decision about what we want for our kids. If we can do that as families, schools or communities, then we can change what our systems are aiming for and how they're pursuing it.

If we see further in our global mission today, it is because we're already standing on the shoulders of giants. Good old Newton. Even an egoist as maniacal as he was could admit he didn't do it alone. If we are to continue to progress, we have to work together, open up knowledge to all, believe that we have the collective intelligence – and solidarity – to solve the problems of the world. Tackling a challenge this complex requires our collective leadership.

Our first step is to create a new blueprint for learning together, throughout our lives, from birth to death, developing our capacity to think anew, do better and take care. Soon, we'll embrace our unending ability to learn, take up coding or data analysis mid-career, become early-childhood teachers or retrain as psychotherapists. We'll access arts classes or sports provision as and when we need them, know our place in the world, pursue contentment over competition. Even economists agree that learning guarantees a return on investment. What's stopping us?

ACKNOWLEDGEMENTS

If it takes a village to raise a child then a small city has helped bring up this book. Hundreds more pages would be needed to adequately thank each person to whom I owe a debt of gratitude. Instead I'd like to dedicate this book to all of the children, teachers and other wonderful people across the world who welcomed me into their classrooms, schools, homes and workplaces, greeting me with warmth, patiently answering my questions and freely sharing their experiences. Every small kindness, helpful introduction and thoughtful comment was deeply appreciated. Today we frequently worry that the world is harsh, divided place. My experience is the opposite. I am filled with hope at the generosity, kindness and hospitality that people met me with everywhere. This book is for you.

Certain individuals showed me particular kindness, opening their doors wide, putting up with endless emails and giving generously of their time and stories. Among those I'd like to thank are Max Haimendorf and Ifrah Khan at KSA, Debbie Penglis, Hannah Barnet and Oli de Botton at School 21, Margy Whalley, Angela Prodger and Rebecca Elliott at Pen Green, Ger Graus at KidZania, Lisa Naylor at P4C, Sugata Mitra, Iroise Dumontheil and Daisy Christodoulou for interviews, Jacob, Sofia, Tor, Eleanor and Dominik Havsteen-Franklin for a grand day out, Sarah-Jayne Blakemore, Toby Greany and Susan Douglas for helpful connections and Joe Kirby for a tour of Michaela School. I'd like to say a special thank you to Jonnie Noakes at Eton College, who hosted me over several visits that shaped my thinking considerably.

In the US, I'm indebted to Brett Schilke at Singularity University, Philipp Schmidt, Mitch Resnick, Kim Smith and Deb Roy at the MIT Media Lab, Diane Tavenner and Sam Strasser at Summit Public Schools, Larry Rosenstock at High Tech High, Kelly Willis and Matthew Needleman at Melrose Elementary, Preston Smith at Rocketship, Peter Croncota, Adie Capaldi, Jeff Li, Dominique Mejia, Allison Holley, Gerard Griffith and the rest of the team at KIPP Infinity for making me feel part of the family, Angela Duckworth, Donald Kamentz, Chad Spurgeon, Sean Talamas and Emily Aisenbrey at the Character Lab for their time and insight, Brent Maddin and Jamey Verrilli at Relay GSE, the kids and staff of North Star Academy, Christina Hinton at Harvard, B. J. Fogg at Stanford, Naya Bloom and Drew Furedi for LA connections and the team at Temple University, Yu Chen, Ying Lin, Jelani Medford, Haley Weaver, Rufan Luo, Brenna Hassinger-Das, Brianna McMillan. Finally, special thanks to Kathy Hirsh-Pasek for inviting me into her home.

In other parts of Europe, I also leaned on the kindness of many wonderful people. In Finland, Anni Rautianen and Saku Tuominen at HundrED, Mervi Kumpulainen and Ilppo Kivivuori at Hiidenkiven School, Pekka Peura in Martinlaakso, Jenna Lähdemäki at Sitra, Olli Vesterinen at the ministry, Aleksi Neuvonen at Demos, as well as the teams at TET 4.0 and MeHackit. In Holland, huge thanks to Christof van Nimwegen. In Paris, a huge thank you to Andreas Schleicher at the OECD, Nicolas Sadirac, Fabienne Haas, Lilas Merbouche and Thomas Guillot at 42 and Francois Taddei at CRI.

I was equally overawed by the generosity of colleagues in Asia. Thank you to Jessica Kehayes and Mingxuan Zhu at the Asia Society, Mingxuan Zhang and Yong Zhao for shaping my understanding, Sophie Chen for numerous connections. In Shanghai, I owe special thanks to Zhang Mingsheng, Stella Shi, Xu Yijie, Xu JinJie and Hailing He and to the students and teachers at YuCai High and Wanhangdu Road Primary. Thanks also to Orestes Za at the Peking University Future School, the Pi-Top team of Jesse Lozano and Ryan Dunwoody, who I met in Beijing. In Hong Kong, my unending gratitude to Joshua Wong and Benny Tai, learners in the truest sense, as well as to Arnett Edwards at UWC and Aisha Spears. In South Korea, the Ashoka and Future Class Networks were unparalleled hosts, generously arranging visits and dinners. Thank you to Seung-Bin Lee, Gwangho Kim, Yumi Jeung and Chanpil Jung at FCN. At Ashoka, Hae-Young Lee, Bo-Kyeong Kwon, Yujin Noh and Jinyee

Ryu went above and beyond, as did Alex Lim, Jinu Baek, Seon Yeong Lee, In-Soo Song and of course Ju-Ho Lee.

Others not featured in the book also helped considerably. In India, Renuka Gupta at Pardada Pardadi and Rukmini Banerji at Pratham were endlessly patient and supportive, even as I failed to visit them after a visa problem. In Mumbai, the FSG team of Samantha King, Gauri Kirtane and Priyamvada Tiwari took me to early childhood centres that weren't featured in the final text. In Singapore, David Hung kindly shared his enormous expertise, while Audrey Jarre and Svenia Busson helped with connections, as did Polly Ackhurst at UWC and Viliana Dzartova at Ashoka.

Going back to where it all began, I'd like to thank my own teachers, along with all the others out there who are doing the most important work. I'd also like to thank the children and staff of Walworth Academy, who bore my early failures with patience, support and a much-needed sense of humour, particularly Faye Kupakuwana, Patricia Toworfe, Mike Higgins and David Jahans in the English department. Thank you also to the students of 10X4 and 10Y1 for inspiring my career and this book, and the debating team for the highs. I'd also like to thank Brett Wigdortz, without whom there'd be no Teach First, and my friend Juliet Cook, for convincing me teaching was the right path and giving me welcome encouragement with the book. To my colleagues at Teach For All thank you for taking up the slack, especially Felicia Cuesta, Kyle Conley, Lucy Ashman and Isy Faingold. Thank you to Steven Farr for many helpful thoughts and introductions and of course to Leigh Kincaid, who has been my partner in crime from day one, and had my corner as I got started. I owe a huge debt of gratitude too to Wendy Kopp, without whom I'd never have made it into the classroom, had the freedom to write, or been able to convince such eminent people around the world to talk to me.

Among those who have played a part in shaping my thinking are Orazio Cappello, Jacob Kestner, Ed Vainker, Rowland Manthorpe, Ed Fornieles and Artur Taevere. Thank you for the many stimulating conversations. Thank you also to Ness Whyte and Bertie Troughton for the titles (look out for *School Runnings* and *Discipline, Discipline, Discipline* in 2019) and Eleanor O'Keeffe and John

Gordon for support with the marketing. I'll also be forever grateful to my elite group of readers Cordelia Jenkins, Matt Lloyd-Rose, Orazio Cappello, Archie Bland, Amol Rajan, Zack Simons, Katie O'Mahoney and Rosie Boycott. Each of them offered valuable insight, sound advice and an honest appraisal of my writing and ideas. I could ask for no greater kindness. The text is vastly better for their input and any weaknesses or errors entirely of my own making.

This book was inspired by Atul Gawande. Following a talk he gave on *Being Mortal* in London three years ago, I told Bea Hemming about a wish I had that there was a similar book about learning. A year later, after reading my proposal, she signed this book to W&N. I'll never forget the faith she showed in me at that stage. I'd also like to thank the rest of the team, Alan Samson, Holly Harley, Cait Davies and Elizabeth Allen for their wit, patience and unerring support. To the wonderful team at L&R, Felicity Rubinstein, Sarah Lutyens, Juliet Mahony and everyone else, a huge thank you for seeing the potential in my early ramblings and backing me to turn them into a book.

The peerless Jenny Lord has been my editor throughout this process. She has been an incredible mentor, patiently poring over countless drafts, always offering the right words of challenge or encouragement, and somehow extracting a passable book from me at the end of it all. I'll never be able to express quite how grateful I am. I'm also hugely lucky that for the past couple of years I've also been able to call my close friend Jane Finigan my agent. Without her support at the earliest stage and tireless advocacy since day one, I would have had neither the confidence nor opportunity to see this through. It means the world.

It would have been still less possible to finish this book without the support of family and friends. Thank you all for not abandoning me as I went into book hiding for two years. A particular thank you to Rosie Boycott and Charlie Howard for putting me up, to Ellis and Ippolita for showing me what a natural born learner looks like and to my brothers Jack, Rory and Max for never letting me forget what I was missing out on and being patient with me as I missed out on all of it. Thank you also to my parents, Mick and Shona Beard. Dad has been my role model throughout my life, showing me what it means to work hard at what you love whilst always finding time for others. His thoughts on early drafts gave me an important boost. My Mum was my first teacher and is still my most important one. She gave me a sense of belonging in the world, instilled in me a lifelong love of reading and kept me going and happy through long months of writing. She is also a teacher by profession, and I became one because of her.

Finally, I thank Daisy, to whom I owe it all. You made this possible, putting up with me when I was there, and when I wasn't. Your insights shaped my thinking, your editing improved my writing and your being brings all the happiness I need. You gave me the inspiration to write, the freedom to do so and the support to make it to the end. Thank you for everything.

FURTHER READING

These are some of the books that have shaped my thinking and influenced my ideas in different ways. I highly recommend them all.

On the Mind

From Bacteria to Bach and Back: The Evolution of Minds, Daniel C. Dennett (2017)

The Knowledge Illusion: Why We Never Think Alone, Steven Sloman and Phillip Fernbach (2017)

The Cultural Origins of Human Cognition, Michael Tomasello (1999)

Language, Cognition and Human Nature: Selected Articles, Steven Pinker (2013)

Phantoms in the Brain: Probing the Mysteries of the Human Mind, Sandra Blakeslee and V. S. Ramachandran (1998)

The Tell-Tale Brain: Unlocking the Mystery of Human Nature, V. S. Ramachandran (2011)

The Future of the Mind, Michio Kaku (2014)

The Learning Brain: Lessons for Education, Sarah-Jayne Blakemore and Uta Frith (2005)

Thinking, Fast and Slow, Daniel Kahneman (2011)

On the Future

The Second Machine Age: Work, Progress, and Prosperity in a Time of Brilliant Technologies, Eric Brynjolfsson and Andrew McAfee (2014)

Sapiens: A Brief History of Humankind, Yuval Harari (2014)

Homo Deus: A Brief History of Tomorrow, Yuval Harari (2017)

Humankind: Solidarity with Nonhuman People, Timothy Morton (2017)

Geek Heresy: Rescuing Social Change from the Cult of Technology, Kentaro Toyama (2015)

In Our Own Image: Will Artificial Intelligence Save Or Destroy Us?, George Zakardakis (2015)

The Singularity is Near: When Humans Transcend Biology, Ray Kurzweil (2005)

On Learning and Schools

Seven Myths about Education, Daisy Christodoulou (2013)

Out of Our Minds: Learning to be Creative, Ken Robinson (2001)

Why Don't Students Like School?: A Cognitive Scientist Answers Questions about How the Mind Works and What It Means for the Classroom, Dan Willingham (2009)

What's the Point of School?: Rediscovering the Heart of Education, Guy Claxton (2008)

Becoming Brilliant: What Science Tells us About Raising Successful Children, Kathy Hirsh-Pasek and Roberta Michnick Golinkoff (2016)

Mind in Society, L. S. Vygotsky (1978)

The Language and Thought of the Child, Jean Piaget (1923)

The Discovery of the Child, Maria Montessori (1909)

The Uses of Literacy, Richard Hoggart (1957)

On Difference

The End of Average: How We Succeed in a World That Values Sameness, Todd Rose (2016)

Far From the Tree: Parents, Children and the Search for Identity, Andrew Solomon (2012)

Neurotribes: The Legacy of Autism and the Future of Neurodiversity, Steve Silberman (2015)

On Performance

The Craftsman, Richard Sennett (2008)

Outliers: The Story of Success, Malcolm Gladwell (2008)

Grit: The Power of Passion and Perseverance, Angela Duckworth, (2016)

Flow, Mihaly Czsikszentmihalyi (1990)

How Children Succeed: Grit, Curiosity and the Hidden Power of Character, Paul Tough (2012)

Originals: How Non-Conformists Move the World, Adam Grant (2016)

On International Examples

Cleverlands: The Secrets Behind the Success of the World's Education Superpowers, Lucy Crehan (2016)

The Smartest Kids in the World: And How They Got That Way, Amanda Ripley (2013)

Who's Afraid of the Big Bad Dragon: Why China Has the Best (and Worst) Education System in the World, Yong Zhao (2015)

Empowering Global Citizens: A World Course, Fernando Reimers (2016)

Little Soldiers: An American Boy, A Chinese School and the Global Race to Achieve, Lenora Chu (2017)

Finnish Lessons 2.0: What Can the World Learn from Educational Change in Finland?, Pasi Sahlberg (2004)

Lessons Learned: How Good Policies Produce Better Schools, Fenton Whelan (2009)

Learning from Singapore: The Power of Paradoxes, Pak Tee Ng (2017)

The Rebirth of Education: Schooling Ain't Learning, Lant Pritchett (2013)

Work Hard, Be Nice: How Two Inspired Teachers Created the Most Promising Schools in America, Jay Matthews (2009)

Learning Reimagined, Graham Brown-Martin (2014)

On Technology

Addiction by Design: Machine Gambling in Las Vegas, Natasha Dow-Schull (2014)

Big Data: A Revolution That Will Transform How We Live, Work, and Think, Kenneth Cukier and Viktor Mayer-Schönberger (2013)

Natural Born Cyborgs: Minds, Technologies and the Future of Human Intelligence, Andy Clark (2004)

Propaganda, Edward Bernays (1928)

Taking Care of the Youth and the Generations, Bernard Stiegler (2010)

The Attention Merchants: The Epic Scramble to Get Inside Our Heads, Tim Wu (2016)

The Shallows: How the Internet Is Changing the Way We Think, Read and Remember, Nicholas Carr (2010)

On Equality and Democracy

Our Kids: The American Dream in Crisis, Robert Putnam (2015)

Respectable: The Experience of Class, Lynsey Hanley (2017)

Miseducation: Inequality, Education and the Working Classes, Diane Reay (2017)

The Braindead Megaphone, George Saunders (2007)

The Spirit Level: Why More Equal Societies Always Do Better, Richard G. Wilson and Kate Pickett (2009)

Rules for Radicals, Saul Alinsky (1971)

Girl Up, Laura Bates (2016)

Better: A Surgeon's Notes on Performance, Atul Gawande (2007)

Being Mortal: Medicine and What Matters in the End, Atul Gawande (2014)

NOTES

Epigraph

1 This line is taken from Flaubert's collected writings. The original reads, 'La vie doit être une éducation incessante; il faut tout apprendre, depuis parler jusqu'à mourir.' The translation is my own and any errors at least partly the fault of my French teachers.

Introduction

1 Pausanias, *Description of Greece*, with an English translation by W. H. S. Jones and H. A. Ormerod, 4 vols., Cambridge, Mass., Harvard University Press, 1918 (accessed at http://www.perseus.tufts.edu/hopper/text?doc=Perseus%3Atext%3A1999.01.0160%3Abook%3D1%3Achapter%3D30%3Asection%3D2).

2 The Allegory of the Cave is featured in Plato's *Republic* in the form of a discussion between Socrates, his mentor, and Glaucon, his brother. It is introduced by Socrates with the intention of showing 'the effect of education and the lack of it on our nature'.

3 Tony Blair visited in 1997, called the inhabitants of the Aylesbury Estate the 'forgotten people', then promptly forgot about them himself.

4 *The Mirror of Literature, Amusement, and Instruction, Volume 5*, John Timbs, 1825, p. 75. Curtis is reported to have used the phrase in a toast.

5 There is no one source of statistics on the number of teachers worldwide. In the UK, approximately 1 in 100 adults are currently teaching. The global proportion is most likely less, so my number may be a little high. UNESCO, however, has forecast that by 2040 the world will need an *extra* 65 million teachers to ensure a quality education for all of the young people predicted to be born in the next 20 years. If even 1 in 1,000 of them bought this book, my publisher would be very happy indeed.

6 UNESCO Institute for Statistics, *More Than Half of Children and Adolescents Are Not Learning Worldwide*, Fact Sheet No. 46, September 2017 (accessed at http://uis.unesco.org/sites/default/files/documents/fs46-more-than-half-children-not-learning-en-2017.pdf).

Part I. Thinking Anew
Chapter 1: Artificial Intelligence

1 www.brettschilke.com

2 Singularity University website, 'About', at https://su.org/about/.

3 Ray Kurzweil, *The Singularity is Near: When Humans Transcend Biology*, New York, Penguin Books, 2006, p. 9.

4 Carl Benedikt Frey and Michael A. Osborne, 'The Future of Employment: How Susceptible are Jobs to Computerisation?', September 2013, Oxford Martin Report, at http://www.oxfordmartin.ox.ac.uk/downloads/academic/The_Future_of_Employment.pdf.

5 Thomas Edison, 1922, cited in Larry Cuban's *Teachers and Machines: The Classroom Use of Technology Since 1920*, New York, Teachers College Press, 1986, p. 9.

6 President Lyndon Johnson made these remarks in an October 1966 speech about the success of an educational television project in accelerating the learning of kids in American Samoa. 'You have recognized that education is the tidal force of our century driving all else ahead of it,' he said, 'and I am told that the pilot program of education that you have started may point the way of learning breakthroughs throughout the Pacific Islands and South-East Asia.'

7 Hubert Dreyfus, *What Computers Can't Do*, New York, Harper & Row, 1963, p. XXXI, at https://archive.org/stream/whatcomputerscan017504mbp/whatcomputerscan017504mbp_djvu.txt.

8 In game two, Deep Blue made a surprising move that Kasparov suspected was human-aided, contravening the rules of the match. IBM refused initially to release the machine's log data, arousing further suspicion. Nothing was ever proved, but the accusation served as a tantalising reminder of the role of human engineers in building, programming and coaching even the most powerful computers.

9 For a great account of the story of human versus computer chess, and of the history of AI in general, try Eric Brynjolfsson and Andrew McAfee, *Race Against the Machine*, Digital Frontier Press, 2011.

10 Gary Kasparov, 'The Chess Master and the Computer', *New York Review of Books,* 11 February 2010, at http://www.nybooks.com/articles/2010/02/11/the-chess-master-and-the-computer/.

11 Ken Jennings, 'My Puny Human Brain', *Slate*, February 2011, at www.slate.com/articles/arts/culturebox/2011/02/my_puny_human_brain.html.

12 Dr. Seuss, *The Lorax*, New York, Random House, 1971.

13 Though this quote is often attributed to American industrialist Henry Ford, especially in classroom posters, there's no hard evidence that he actually said it. According to quoteinvestigator.com, the earliest recorded attribution comes from a 1947 edition of *Reader's Digest*, which published the quotation in a freestanding, unattributed format as follows, 'Whether you believe you can do a thing or not, you are right.'

14 note 14 text TK.

15 Sugata Mitra, 'Let's Build a School in the Cloud', TED talk, 3 May 2013.

16 Ibid.

17 *Los Angeles Times*, at e.g. http://www.latimes.com/local/lanow/la-me-ln-la-unified-ipad-settlement-20150925-story.html. It should be noted that the then Superintendent insisted that 'nothing was done in any inappropriate way whatsoever', and Pearson said that the bidding process was separate and understood by them to be open and competitive to all.

18 'Let Them Eat Tablets', at http://www.economist.com/node/21556940.

19 Organization for Economic Co-operation and Development, 'Students, Computers and Learning', Paris, 2015.

20 See, for example, Jane Wakefield, 'Foxconn replaces 60,000 factory workers with robots', BBC website, 25 May 2016.

21 Kasparov, at http://www.nybooks.com/articles/2010/02/11/the-chess-master-and-the-computer/.

22 John Lanchester, 'The Robots are Coming and They're Going to Eat All the Jobs', *London Review of Books*, vol. 37, no. 5 (5 March 2015), pp. 3–8.

23 In Eric Brynjolfsson and Andrew McAfee, *Race Against the Machine*, Digital Frontier Press, 2011, there's a useful introductory section on 'The Paradox of Robotic Progress'.

24 Philip K. Dick, 'Minority Report', in *Minority Report*, London, Gollancz, 2002. I am sad that I mention Tom Cruise here rather than the sci-fi writer.

25 Andrew Griffiths, 'How Paro the robot seal is being used to help UK dementia patients', *Guardian*, 8 July 2014.

26 The Unschool movement is the

twenty-first-century version of homeschooling, which is becoming increasingly popular worldwide, particularly among the know-better-than-yous of Silicon Valley.

27 For more on AltSchool see for example, 'Inside the School Silicon Valley Thinks Will Save Education', *Wired*, 4 May 2015, or 'AltSchool's Disrupted Education', *New Yorker*, 7 March 2016.

28 I spent a day at AltSchool while I was in the Valley. Having read gushing articles in *Wired* and the *New Yorker*, I went expecting to be dazzled and was a little disappointed by the reality. Yes, the model of the shopfront schoolroom with the backroom software team was sexy and exciting – one developer zipped across the office on an electric scooter – but the classrooms felt, well, like any other OK classrooms. The kids were the daughters and sons of affluent San Fransiscans, destined to excel in any learning environment and determinedly individual, like *X-Men First Class*, I noted, without the superpowers.

29 Robert D. Putnam, *Our Kids: The American Dream in Crisis*, New York, Simon & Schuster, 2015.

Chapter 2: Born to Learn

1 Jonathon Roy, 'The Power of Babble', *Wired*, 1 April 2007, at https://www.wired.com/2007/04/truman/.

2 Ibid.

3 Ibid.

4 From Deb Roy et al., 'The Human Speechome Project', presented at the 28th Annual Conference of the Cognitive Science Society, July 2006, at https://www.media.mit.edu/cogmac/publications/cogsci06.pdf.

5 Betty Hart and Todd R. Risley, 'The Early Catastrophe: The 30 Million Word Gap', *American Educator*, vol. 27, no. 1 (Spring 2003), pp. 4–9, at https://www.aft.org/sites/default/files/periodicals/TheEarlyCatastrophe.pdf.

6 Ibid.

7 Anne Fernald, Virginia A. Marchman and Adriana Weisle, 'SES differences in language processing skill and vocabulary are evident at 18 months', *Dev Sci*, 16: 234–248, 2013

8 Roberta Michnick Golinkoff and Kathy Hirsh-Pasek, *Becoming Brilliant: What Science Tells Us About Raising Successful Children*, American Psychological Association, 2016 (p. 8)

9 J. R. Saffran, R. N.; Aslin and, E. L. Newport (1996), 'Statistical Learning by 8-Month-Old Infants', *Science*, vol. 274, no. 5294, pp.): 1926–1928.

10 Alison Gopnik, Andrew N. Meltzoff and Patricia K. Kuhl, *The Scientist in the Crib: Minds, Brains, and How Children Learn*, New York, William Morrow & Co., 1999. Published in the UK as *How Babies Think: The Science of Childhood*, London, Weidenfeld & Nicolson, 1999.

11 Eino Partanen, Teija Kujalaa, Risto Näätänen, Auli Liitolaa, Anke Sambethf, and Minna Huotilainena, 'Learning-induced neural plasticity of speech processing before birth', PNAS 2013, 110 (37) 15145–15150.

12 Michael Tomasello, *The Cultural Origins of Human Cognition*, London, Harvard University Press, 2003, pp. 1–12.

13 Yuval Harari's *Sapiens: A Brief History of Humankind*, London, Vintage, 2015, gives a highly readable history of our cognitive revolution.

14 Ibid, Michael Tomasello, p. 7.

15 Ibid, Michael Tomasello, pp. 13–55.

16 Coronado, N. (2013, November 19). The critical period hypothesis on language acquisition studied through feral children. Retrieved from http://www.newsactivist.com/en/articles/knowledge-media/critical-period-hypothesis-language-acquisition-studied-through-feral.

17 Harry F. Harlow, 'The nature of love', *American Psychologist*, 13, 673–685, 1958.

18 See for example, John Bowlby, *Attachment and Loss: Vol. 1. Loss*. New York, Basic Books, 1969.

19 These findings, from research led

by Leslie Seltzer at the University of Wisconsin-Madison's Child Emotion Lab, were reported by Katherine Harmon in *Scientific American* in May 2010.

20 Sue Gerhardt, *Why Love Matters: How Affection Shapes a Baby's Brain*, London, Taylor & Francis, 2nd edn, 2015.

21 As Michael Meaney and all report in 'Epigenetic programming by maternal behaviour', *Nature Neuroscience* 7, 847–854, 2004, a process called methylation meant proteins were released in the rat-pups' brains in response to environmental stimulation (the mother's licks), causing their DNA to unravel to a greater or lesser extent fixing changes in the rat pup brains that would endure in future generations.

22 UCLA professor Dan Siegel calls this 'mindsight' good paper to read.

23 *Seven Up* and *Child of Our Time* were two British TV shows that followed the lives of a group of average kids over the course of several decades.

24 Lawrence J. Schweinhart, *The High/Scope Perry Preschool Study Through Age 40: Summary, Conclusions, and Frequently Asked Questions*, High/Scope Press, 2004.

25 James Heckman at http://bostonreview.net/forum/promoting-social-mobility-james-heckman. For more on the curve visit: https://heckmanequation.org/resource/the-heckman-curve/.

26 Patricia K. Kuhl, Feng-Ming Tsao, and Huei-Mei Liu, 'Foreign-language experience in infancy: Effects of short-term exposure and social interaction on phonetic learning', PNAS 2003 100 (15) 9096–9101.

27 Maia Szalavitz, 'Like Crack for Babies: Kids Love Baby Einstein, But They Don't Learn from It', *Time*, 7 September 2010.

28 Ibid.

29 For more on this, see Erica Christakis's *The Importance of Being Little: What Preschoolers Really Need from Grownups*, New York, Viking Press, 2016.

30 Daphna Bassok, Scott Latham, Anna Rorem, 'Is Kindergarten the New First Grade?', *AERA Open*, at https://doi.org/10.1177/2332858415616358, January, 2016.

31 David Whiteread at http://www.cam.ac.uk/research/discussion/school-starting-age-the-evidence.

32 Ibid.

33 Thomas S. Dee, Hans Henrik Sievertsen, 'The Gift of Time? School Starting Age and Mental Health', *NBER Working Paper No. 21610*, October 2015.

34 *We Live in Public* (2009) is a brilliant documentary directed by Ondi Timoner about an internet millionaire, Josh Harris, who decided to film his whole existence.

35 This is brilliantly reported on in Jon Ronson's *The Psychopath Test: A Journey through the Madness Industry*, London, Picador, 2011.

36 Helen Keller, *The Story of My Life,* 1903.

37 T. S. Eliot, from 'Little Gidding', in *The Four Quartets*, London, Faber, 1944.

Chapter 3: Brain Gains

1 Francis Galton, *Hereditary Genius: An Inquiry into Its Laws and Consequences*, London, Macmillan, 1869.

2 Todd Rose gives a great in-depth account of the history of intelligence in *The End of Average: How We Succeed in a World That Values Sameness*, London, Penguin Books, 2017.

3 Eric R. Kandel, *In Search of Memory: The Emergence of a New Science of Mind*, New York, W. W. Norton & Co., 2006. Kandel also comes up in Nicolas Carr's *The Shallows: How the Way the Internet is Changing the Way We Think, Read and Remember*, London, Atlantic, 2010, and Andrew Solomon's *Far from the Tree: Parents, Children and the Search for Identity*, London, Scribner, 2012, both of which are brilliant – in very different ways – on human intelligence.

4 This is an image used by Proust in *In Search of Lost Time*. I was always baffled by the English translation, 'the little Japanese water flowers'. Then I saw one of my wife's godchildren playing with one of these toys and suddenly it all made sense.

5 There is a great debate on this between on the one hand those like Robert Plomin, a psychologist who says that it is clear that intelligence is heritable, and on the other Oliver James, who says it's all down to nurture.

6 Peter Diamandis, 'The Way We Learn Today is Just Wrong', *Huffington Post*, 19 June 2016, at https://www.huffingtonpost.com/entry/the-way-we-learn-today-is-just-wrong_us_5766c8c9e4b0092652d7a173.

7 In a 1971 article 'Designing Organizations for an Information-Rich World', artificial intelligence pioneer Herbert Simon explained it like this, 'in an information-rich world, the wealth of information means a dearth of something else: a scarcity of whatever it is that information consumes. What information consumes is rather obvious: it consumes the attention of its recipients. Hence a wealth of information creates a poverty of attention and a need to allocate that attention efficiently among the overabundance of information sources that might consume it.' In: Martin Greenberger, *Computers, Communication, and the Public Interest*, Baltimore. MD: The Johns Hopkins Press. pp. 40–41.

8 Jacob Weisberg references these statistics in his article, 'We Are Hopelessly Hooked', *New York Review of Books*, 25 February 2016. I am deeply indebted to the piece, as I am to Ian Leslie's 'The Scientists Who Make Addictive Apps', *1843*, October 2016, for fuelling my interest in the irresistibly murky field of captology.

9 B. J. Fogg and Clifford Nass, 'Silicon Sycophants: The Effects of Computers That Flatter', *International Journal of Human–Computer Studies*, vol. 46 (1997), pp. 551–61.

10 For a great account of the work of B. F. Skinner, read *X, Y, Z*.

11 This means their families earn under £16,190 per year.

12 Steven Pinker, *The Blank Slate: The Modern Denial of Human Nature*, London, Penguin, 2002.

13 Ibid, Eric Kandel.

14 Gerald Eugene Myers, *William James: His Life and Thought*, New York, Yale University Press, 1986, p. 204.

15 Daisy Christodoulou, *Seven Myths about Education*, Abingdon, Routledge, 2014.

16 For a deeper exploration of why a knowledge-rich education still matters, read Christodoulou's *Seven Myths*. It's a brilliant take on the subject of knowledge, memorisation and cultural literacy, showing why we need to *know* in order to be able to *think*.

17 Daniel Willingham, *Why Don't Students Like School?*, San Francisco, Jossey-Bass, 2009, p. 43.

18 Ibid., p. 48.

19 Christof van Nimwegen, 'The Paradox of the Guided User: Assistance Can be Counter-Effective', doctoral thesis, University of Utrecht, 2008, at https://dspace.library.uu.nl/handle/1874/26875.

20 Peter Blatchford, 'Reassessing the Impact of Teaching Assistants', London, Routledge, 2012.

21 Daniel Dennett, *From Bacteria to Bach and Back: The Evolution of Minds*, London, Allen Lane, 2017.

22 Lisanne Bainbridge, 'Ironies of Automation', *Automatica*, vol. 19, no. 6 (November 1983), pp. 775–9, and also at http://www.bainbrdg.demon.co.uk/Papers/Ironies.html.

23 E. L. Bjork and R. A. Bjork, 'Making Things Hard on Yourself, But in a Good Way: Creating Desirable Difficulties to Enhance Learning', in M. A. Gernsbacher and J. Pomerantz (eds.), *Psychology and the Real World: Essays Illustrating Fundamental Contributions to Society*, 2nd edn, New York, Worth, 2014, pp. 59–68,

and at https://teaching.yale-nus.edu.sg/
wp-content/uploads/sites/25/2016/02/
Making-Things-Hard-on-Yourself-but-
in-a-Good-Way-20111.pdf.

24 Eleanor Maguire, 'London taxi drivers
and bus drivers: a structural MRI
and neuropsychological analysis',
at https://www.ncbi.nlm.nih.gov/
pubmed/17024677 studies of cabbies'
brains.

25 Stanford Persuasive Tech Lab
homepage at http://captology.stanford.
edu/ (accessed on 15 November 2017).

26 Jacob Weisberg, New York Review of
Books, ibid.

27 Natasha Dow-Schüll, Addiction by
Design: Machine Gambling in Las
Vegas, Princeton, Princeton University
Press, 2012.

28 Ibid., p. 167.

29 Gerald Maurice Edelman and Giulio
Tononi, A Universe of Consciousness:
How Matter Becomes Imagination,
London, Penguin, 2000.

Part II. Doing Better

Chapter 4: Just Do It

1 Time for Change: An Assessment
of Government Policies on Social
Mobility 1992–2017, Social
Mobility Commission, June
2017, at https://www.gov.uk/
government/uploads/system/uploads/
attachment_data/file/622214/
Time_for_Change_report_-_An_
assessement_of_government_policies_
on_social_mobility_1997-2017.pdf.

2 Pascual Restrepo, 'Skill Mismatch
and Structural Unemployment', 2015,
at http://pascual.scripts.mit.edu/
research/01/PR_jmp.pdf.

3 Carl Benedikt Frey and Michael
A. Osborne, 'The Future of
Employment: How Susceptible are
Jobs to Computerisation?', September
2013, Oxford Martin Report, at
http://www.oxfordmartin.ox.ac.uk/
downloads/academic/The_Future_of_
Employment.pdf.

4 Joel Mokyr, Chris Vickers, Nicolas
L. Ziebarth, 'The History of
Technological Anxiety and the Future
of Economic Growth: Is This Time
Different?, Journal of Economic
Perspectives, Vol. 29 No.3 Summer
2015.

5 As identified by Burning Glass
Technologies – a company delivering
'job market analytics that empower
employers, workers, and educators to
make data-driven decisions'.

6 Ibid., John Lanchester.

7 Douglas Adams must bear some
responsibility for the way our world
is headed. His Hitchhiker's Guide
to the Galaxy has surely fired the
imaginations of every AI expert
feverishly working to make humanity
obsolete.

8 So must J. R. R. Tolkein.

9 And George Lucas.

10 Scott Sayare, 'A Computer Academy in
France Defies Conventional Wisdom',
New York Times, 15 November
2013, at http://www.nytimes.
com/2013/11/16/world/europe/
in-france-new-tech-academy-defies-
conventional-wisdom.html.

11 Xavier Niel, 'La Philosophie 42', at
http://www.42.fr/ledito/.

12 Ibid., Scott Sayare.

13 http://www.kidzania.com/the-
company.html.

14 Rebecca Mead, 'A City Run by
Children', New Yorker, 19 January
2015.

15 Education Action Zones were a
relatively short-lived government
policy in the late 1990s, aimed
at raising the attainment of
disadvantaged students through
partnering groups of schools with
local businesses in the hope that they'd
inspire innovation. A small positive
effect was seen at primary level.

16 Carol S. Dweck, Mindset: Changing
the Way You Think to Fulfil Your
Potential, rev. edn, London, Robinson,
2017.

17 I spent valuable time at Eton during
the research for this book, most
memorably attending a staff meeting
in a hall decorated with the busts of
former British prime ministers. The

school has given the UK 17 of its 53 prime ministers, the most recent being David Cameron, while among the names scratched into the wood-panelled walls of the meeting room was that of the poet Percy Shelley, who had also been a pupil at the school. It was that type of place.

Beyond the school's adherence to the highest academic standards, I was above all struck by the commitment of the staff to creating the culture and opportunities that enabled pupils to find their passion and establish their place in the school community. In accepting kids to the school, they didn't accept only the brightest, but those that might add something to the life of the school, whether through leadership, creativity, music, drama, sports or something else.

The boys – and they are all boys – were responsible for a huge proportion of their own time, and used much of it in the establishment and running of societies. In these, they would invite global or national experts in an area of particular interest to them to come to meet them for dinner at the school. Many accepted these invitations, from the Israeli ambassador to the UK to the chief executive of Heathrow Airport. Indeed, the school was almost better known for the success of its extra-curricular activities. The actors Damian Lewis, Dominic West and Eddie Redmayne all attended the school.

18 Eric Schmidt, 'How Google Manages Talent', *Harvard Business Review*, September 2014, at https://hbr.org/2014/09/how-google-manages-talent.

19 Harriet Agnew, '"Big Four" look beyond academics', *Financial Times*, 28 January 2016, at https://www.ft.com/content/b8c66e50-beda-11e5-9fdb-87b8d15baec2.

20 The School of Life is a global organisation that promotes emotional intelligence, applying 'psychology, philosophy and culture to everyday life'. It has noble aims and a hefty price tag for its courses.

21 Lifelong learning credits in Singapore as reported by the *Straits Times*, at http://www.straitstimes.com/singapore/education/starting-jan-1-singaporeans-aged-25-and-above-will-get-500-credit-to-upgrade.

22 'Graduates earn £500,000 more than non-graduates', *Telegraph,* 16 July 2015, at http://www.telegraph.co.uk/finance/jobs/11744118/Graduates-earn-500000-more-than-non-graduates.html.

23 Paul Oyer, *Everything I Ever Needed to Know about Economics I Learned from Online Dating*, Harvard Business Review Press, p. 178.

Chapter 5: Creation

1 There are a ton of these. For example: 'The struggle of maturity is to recover the seriousness of the child at play' – Friedrich Nietzsche; 'We have to continually be jumping off cliffs and developing our wings on the way down' – Kurt Vonnegut.

2 Maria Montessori, *Discovery of the Child*, 1903.

3 You can see these in more detail at http://iwantyoutowantme.org/ and http://wefeelfine.org/.

4 For a good account of John Searle's Chinese room experiment, try Searle's own 1980 paper, 'Minds, Brains and Programs', *Behavioral and Brain Sciences*, vol. 3, no. 3 (September 1980), pp. 417–24.

5 Benjamin Bloom (ed.), *Developing Talent in Young People*, New York, Ballantine Books, 1985.

6 Malcolm Gladwell, *Outliers: The Story of Success*, London, Penguin Books, 2008, is deeply concerned with learning, as well as an inevitably brilliant read.

7 Adam Grant, *Originals: How Non-conformists Change the World*, London, Ebury, 2016, p. 9.

8 In a piece for the *New York Times,* 'How to Raise a Creative Child. Step One: Back Off', 30 January 2016, Adam Grant references a 1989 study

by John S. Dacey, 'Discriminating Characteristics of the Families of Highly Creative Adolescents', *Journal of Creative Behaviour*, Volume 23, Issue 4, December 1989, pp. 263–271.

9 For a great account of this, read Jonah Lehrer's *Imagine*, if you can get a copy.

10 Ibid., Adam Grant, *New York Times*.

11 Ibid.

12 Ibid.

13 Ibid.

14 Carol Dweck, *Mindset: Changing the Way You Think to Fulfil Your Potential*, London, Robinson, 2017.

15 K. Anders Ericsson, Ralf Th. Krampe and Clemens Tesch-Römer, 'The Role of Deliberate Practice in the Acquisition of Expert Performance', *Psychological Review*, vol. 100, no. 3 (1993), pp. 363–406, at http://projects.ict.usc.edu/itw/gel/EricssonDeliberatePracticePR93.pdf.

16 They have been challenged by counter-studies, which claim that the importance of deliberate practice has been overstated, explaining an 18 per cent variation in elite-level performance. Other factors include environmental conditions or genetic inheritance. Even the ability to practise deliberately has been found to be inheritable in some cases.

17 Rebecca Jones, 'Entries to arts subjects taken at Key Stage 4', Education Policy Institute at https://epi.org.uk/report/entries-arts-subjects/.

18 'Schools Minister Makes No Apology for Sidelining the Arts', at http://www.ahsw.org.uk/news.aspx?id=1540

19 Janet Murray, 'Tony Blair's Advisor Starts a Free School', *Guardian*, 3 January 2012, at https://www.theguardian.com/education/2012/jan/03/tony-blair-adviser-starts-free-school.

20 It means 'unique'. I googled it.

21 OCED, 'The Survey of Adult Skills', *Readers Companion*, 2nd edn, Paris, 2013, as reported for example by Randeep Ramesh, 'England's young people near bottom of global league table for basic skills', *Guardian*, 8 October 2013.

22 Richard Sennett, *The Craftsman*, London, Allen Lane, 2008, p. 105.

23 Marshall McLuhan, *Understanding Media: The Extensions of Man*, New York, Mentor, 1964.

24 I ran out of steam after twelve. Looking at the website, I realised I'd failed because I only thought of creating a two-dimensional model, when of course you could combine the bricks in three dimensions.

25 Alain Badiou (trans. Peter Bush), *In Praise of Love*, London, Serpent's Tail, 2012.

Chapter 6: Class Act

1 This scene is taken from a video produced by the Future Class Network (see end of chapter) about 'Flipping the Classroom'.

2 I first encountered this story in Atul Gawande's *Better: A Surgeon's Notes on Performance* (London, Profile, 2007), another brilliant book that everyone should read.

3 Ignaz Semmelweis (trans. K. Codell Carter), *Etiology, Concept and Prophylaxis of Childbed Fever* [1861], Madison, University of Wisconsin Press, 1983, pp. 1–49.

4 Howard Gardner, *Frames of Mind: Theory of Multiple Intelligences*, 2nd edn, London, Fontana, 1983, -68a handful e, ogle" Yeager2Studying Self-Control'some of the lowest levels of literacy in the developed world.e things down.

5 Doug Lemov, *Teach Like a Champion: 49 Techniques That Put Students on the Path to Knowledge*, San Francisco, Jossey-Bass, 2010.

6 Ian Leslie, 'The revolution that's changing the way your child is taught', *Guardian*, 11 March 2015.

7 Ibid.

8 Doug Lemov at teachlikeachampion.com/.

9 Benjamin S. Bloom, 'The 2 Sigma Problem: The Search for Methods of Group Instruction as Effective as One-to-One Tutoring', *Educational*

Researcher, vol. 13, no. 6 (June–July 1984), pp. 4–16.

10 John Hattie, *Visible Learning: A Synthesis of over 800 Meta-Analyses Relating to Achievement*, London, Routledge, 2009.

11 Charles Duhigg, 'What Google Learned From Its Quest to Build the Perfect Team', *New York Times*, 25 February 2015.

12 Ibid., John Hattie.

13 Playing to the Gallery: Grayson Perry's talk at the Royal Festival Hall, 16 September 2014.

14 See, among other papers, Steven G. Rivkin, Eric A. Hanushek, and John F. Kain, 'Teachers, Schools, and Academic Achievement', *Econometrica*, vol. 73, no. 2, March 2005, p. 417–458.

15 Dylan Wiliam, 'Assessment for Learning: why, what and how', Institute of Education, edited transcript of a talk given at the Cambridge Assessment Network Conference on 15 September 2006 at the Faculty of Education, University of Cambridge.

16 Eric Hanushek, 'The economic value of higher teacher quality', *Economics of Education Review*, 30, 2011, pp. 466–479.

17 The New Teacher Project, 'The Mirage: Confronting the Hard Truth About Our Quest for Teacher Development', 2015.

18 Anders Ericsson, 'The Making of an Expert', *Harvard Business Review*, July 2007.

19 Ibid., John Hattie.

Part III. Taking Care

Chapter 7: Big Data

1 Amanda Ripley, 'The World's Schoolmaster', *Atlantic*, July 2017.

2 The OECD website has a host of sample PISA questions that you can try out at home, at http://www.oecd.org/pisa/pisaproducts/pisa-test-questions.htm. I expect you to score full marks.

3 'On the World Stage U.S. Students Fall Behind', *Washington Post*, 6 December 2016.

4 Ibid., Amanda Ripley.

5 Michael Gove, 'The benchmark for excellence: Can British schools catch up with other nations?', *Independent*, 6 January 2011.

6 Ju-Ho Lee, Hyeok Jeong, Song-Chang Hong, 'Is Korea Number One in Human Capital Accumulation?: Education Bubble Formation and Its Labor Market Evidence', *KDI School of Pub Policy & Management Paper* No. 14-03, 9 August 2014.

7 Michel Foucault, *Discipline and Punish, Part Three, the Means of Correct Training* (1995), Discipline & Punish: The birth of the prison. [Trans. A. Sheridan, 1977.]. New York, Vintage Books, p. 184.

8 The answer is (5).

9 Simon Mundy, 'South Korea's Millionaire Tutors', *Financial Times*, 16 June 2014.

10 Sung-Wan Kim and Jin-Sang Yoon, 'Suicide, an Urgent Health Issue in Korea', *Journal of Korean Medical Science*, 2013 Mar; 28(3), pp. 345–347.

11 I first read this story in Amanda Ripley's excellent *The Smartest Kids in the World*, New York, Simon & Schuster, 2013. There are multiple online news reports recounting further details of the case.

12 Michael Horn, 'Meister Of Korean School Reform: A Conversation With Lee Ju-Ho', *Forbes*, 14 March 2014.

13 Milena Mikael-Debass, 'Land of the Robots: Why South Korea has the highest concentration of robots in the world', 24 May 2017, at https://news.vice.com/story/south-korea-has-the-most-robot-workers-per-human-employee-in-the-world.

14 See 'PISA 2012 Results in Focus' and 'PISA 2009 Results: Executive Summary' at https://www.oecd.org/pisa/keyfindings/pisa-2012-results-overview.pdf and https://www.oecd.org/pisa/pisaproducts/46619703.pdf.

15 This is written down first by Guan Zhong, a seventh-century Qi Dynasty

ruler, who governed in the Confucian tradition.

16 Yong Zhao, *Who's Afraid of the Big Bad Dragon: Why China Has the Best (and Worst) Education System in the World*, San Francisco, Jossey-Bass, 2014.

17 Yong Zhao, *Who's Afraid of the Big Bad Dragon*, Introduction.

18 Yuval Noah Harari, 'Yuval Noah Harari on big data, Google and the end of free will', *Financial Times*, 26 August 2016.

19 'China Invents the digital totalitarian state', *The Economist*, 17 December 2016.

20 Daniel Koretz, *The Testing Charade: Pretending to Make Schools Better*, London, University of Chicago Press, 2017.

Chapter 8: True Grit

1 This section draws on Jay Mathews, *Work Hard, Be Nice: How Two Inspired Teachers Created the Most Promising Schools in America*, Chapel Hill, NC, Algonquin Books, 2009; and Paul Tough, *How Children Succeed: Grit, Curiosity, and the Hidden Power of Character*, London, Random House, 2013, as well as a phone interview with Dave Levin.

2 'Mike & Dave on Oprah', April 2006 (accessed at vimeo.com/91438778).

3 Ibid.

4 KIPP segment on *60 Minutes,* August 2000, at https://vimeo.com/91447154.

5 Malcolm Gladwell wrote about them in his book *Outliers: The Story of Success*, London, Penguin Books, 2008.

6 Sarah Montague, 'Character Lessons', *The Educators*, BBC Radio 4, 30 May 2016

7 See example Seligman speech on YouTube.

8 Martin Seligman, *Learned Optimism: How to Change Your Mind and Your Life,* New York, Random House, 2006, Chapter 1. iii. Depression.

9 Office for National Statistics, 'Young people's well-being:

2017', at https://www.ons.gov.uk/ peoplepopulationandcommunity/ wellbeing/articles/youngpeoples wellbeingandpersonalfinance/2017.

10 See for example, 'Investing in Mental Health', a 2003 report by the World Health Organization, at http:// www.who.int/mental_health/media/ investing_mnh.pdf. In 2017 the WHO also published 'Depression and Other Common Mental Disorders', which estimated that 300 million people worldwide suffer from depression today.

11 Avner Offer, cited in Jen Lexmond and Richard Reeves, 'Parents are the principal architects of a fairer society: Building Character', London, Demos, p. 23.

12 Angela Duckworth, *Grit: The Power of Passion and Perseverance*, London, Vermilion, 2016 (pp. 1–14), which is the source for much material in this section.

13 Ibid., p. 15.

14 Ibid., p. 8.

15 Ibid., pp. 15–34.

16 Ibid. Sarah Montague.

17 Marty Seligman and Christopher Peterson, *Character Strengths and Virtues: A Handbook and Classification*, New York, American Psychological Association and Oxford University Press, 2004.

18 Good accounts of the Marshmallow Test have been written by Jonah Lehrer, 'Don't', *New Yorker*, 18 May 2009; Maria Konnikova, 'The Struggles of a Psychologist Studying Self-Control', *New Yorker*, 9 October 2014; and Michael Bourne, 'We Didn't Eat the Marshmallow. The Marshmallow Ate Us', *New York Times*, 10 January 2014.

19 Cited in ibid, Jonah Lehrer.

20 Cited in ibid, Jonah Lehrer.

21 See for example, Steven M. Brunwasser, Jane E. Gillham, and Eric S. Kim, 'A Meta-Analytic Review of the Penn Resiliency Program's Effect on Depressive Symptoms', Journal of Consulting and Clinical Psychology. Dec 2009; 77(6): 1042–1054.

22 These computerised tests are designed to help teachers and parents 'improve learning for all students and make informed decisions to promote a child's academic growth'. They're also, as Seung-Bin's travails in Seoul indicate, a source of stress for students.

23 Emma Young, 'Iceland knows how to stop teen substance abuse but the rest of the world isn't listening', *Mosaic*, 17 January 2017.

24 David Paunesku, Gregory M. Walton, Carissa Romero, Eric N. Smith, David S. Yeager, and Carol S. Dweck, 'Mind-Set Interventions Are a Scalable Treatment for Academic Underachievement', Psychological Science OnlineFirst, published on 10 April, 2015 as doi:10.1177/0956797615571017.

25 Daniel Pink's *Drive: The Surprising Truth about What Motivates Us*, Edinburgh, Canongate, 2011, is a brilliant book taking a fresh look at motivation. Rather than using classic extrinsic techniques like rewards and punishments, he argues instead that motivation is at its best when it's *intrinsic*. You are most motivated in activities where you experience autonomy, mastery and purpose.

Chapter 9: Mind Control

1 This line is taken from Katherine Boo, *Behind the Beautiful Forevers*, London, Portobello, 2014.

2 According to his lawyer, Michael Vidler. 'He's so young but so wise that you can't help but have a lot of time for him . . . He is every mother's son – filial, polite, principled, hard-working.' Vidler described Wong's parents, Grace and Roger, as 'a very quiet, middle-class, ordinary family', rather than activists.

3 Joshua Wong, 'Scholarism on the March', *New Left Review*, 92, March–April 2015, at https://newleftreview.org/II/92/joshua-wong-scholarism-on-the-march.

4 *Joshua: Teenager v. Superpower* is the title of a 2017 Netflix documentary.

5 Climbing the ziggurat is an image borrowed from Tom Wolfe's *The Right Stuff*.

6 Ibid., Joshua Wong.

7 Haruki Murakami, in his Jerusalem Prize acceptance speech, quoted in the *Jerusalem Post*, 15 February 2009.

8 For example by Google's Eric Schimdt at the Techonomy Conference in August 2010.

9 You can argue that this all started with the Enlightenment. Martin Luther with his vernacular Bibles and René Descartes with his *cogito ergo sum* were forerunners of the internet era, freeing people from a Church-enforced obscurantism that had stymied learning for a thousand years. Along with Francis Bacon, they aimed for the first time to root knowledge in individual experience and observable phenomena, rather than higher authority. It brought about the democratisation of knowledge, but it also opened the door to 'People in this country have had enough of experts' and 'I love the poorly educated.'

10 Sam Wineburg, 'Evaluating Information: The Cornerstone of Civic Online Reasoning', November 2016, at https://sheg.stanford.edu/upload/V3LessonPlans/Executive%20Summary%2011.21.16.pdf.

11 Harroon Siddique, 'Teach schoolchildren how to spot fake news, says OECD', *Guardian*, 18 March, 2017.

12 Hannah Arendt, *The Origins of Totalitarianism*, 1951.

13 Ben Quinn, 'Nursery raised fears of radicalisation over boy's cucumber drawing', *Guardian*, 11 March 2016.

14 All from Maajid Nawaz, *Radical: My Journey from Islamist Extremism to a Democratic Awakening*, London, W. H. Allen, 2012, pp. 30–68.

15 Jon Ronson, *Them: Adventures With Extremists*, London, Picador, 2001.

16 Leading neuroscientists such as Professor Sarah-Jayne Blakemore at UCL in London have shown that the brain goes through a further 'sensitive period' during adolescence during

which it is particularly susceptible to the social influence of peers. This is a double-edged sword, meaning that our brains remain open to learning throughout our youth and young adulthood, but also meaning that they are potentially deeply influenced by peer pressure, whether good or bad.

17 Steven Sloman and Philip Fernbach, *The Knowledge Illusion: Why We Never Think Alone*, London, Macmillan, 2017, p. 8.

18 http://www.newyorker.com/magazine/2017/02/27/why-facts-dont-change-our-minds.

19 Daniel Kahneman, *Thinking, Fast and Slow*, New York, Farrar, Straus and Giroux, 2011.

20 Alexander Pope, in 'An Essay on Criticism', 1709.

21 I also like Francis Bacon's take on this: 'a little philosophy inclineth man's mind to atheism', he wrote, 'but depth in philosophy bringeth men's minds about to religion', from 'Of Atheism', in *Meditationes sacrae*,1597.

22 Ibid., Steven Sloman and Philip Fernbach, p. 5.

23 Bruno Latour, 'Why Has Critique Run out of Steam? From Matters of Fact to Matters of Concern', at http://www.bruno-latour.fr/sites/default/files/89-CRITICAL-INQUIRY-GB.pdf.

24 Ibid. Haruki Murakami.

25 Edward Bernays, *Propaganda*, 1928.

26 In a great Noam Chomsky talk on education that you can watch online, the old leftist tells an intriguing tale of a reactionary response to the free lovin' 1960s. Shaken by what they perceived to be the excessive freedom and independence of the beat generation, particularly among minority groups, two shadowy cabals of businessmen published documents in the mid-1970s urging that the institutions of the state, including the schools, should be used to grow a more pliant population. *The Powell Memo*, written by the lawyer who'd defended the rights of Big Tobacco companies to keep advertising cigarettes, advocated continual surveillance of textbook content, while *The Crisis of Democracy* was a report on 'the governability of democracies' that recommended the use of education to stabilise social and political hierarchies.

27 Duff McDonald, *The Golden Passport: Harvard Business School, the Limits of Capitalism, and the Moral Failure of the MBA Elite*, New York, HarperCollins, 2017.

28 Tim Wu, *The Attention Merchants: The Epic Struggle to Get inside Our Heads*, London, Atlantic, 2017.

29 It was Graham Brown-Martin at a December 2013 conference hosted by the European Commission, on education in the digital era.

30 Wikipedia explains that 'The MacArthur Fellows Program, MacArthur Fellowship, or "Genius Grant", is a prize awarded annually by the John D. and Catherine T. MacArthur Foundation typically to between 20 and 30 individuals, working in any field, who have shown "extraordinary originality and dedication in their creative pursuits and a marked capacity for self-direction" and are citizens or residents of the United States.'

31 George Saunders, 'The Braindead Megaphone', in *The Braindead Megaphone: Essays*, New York, Riverhead, 2007, pp. 203–204, 'The United States of Huck'.

32 Hattie, *Visible Learning*.

33 From the Netflix documentary *Joshua: Teenager v. Superpower* (2017).

34 Saunders, 'The Braindead Megaphone', p. 55. This is taken from 'The New Mecca', which I guarantee is the best thing you'll ever read about Dubai.

Part IV. Coda

Chapter 10: Fail Better

1 *The New Urban High School: A Practitioner's Guide*, Cambridge, Mass., Big Picture Co., 1998.

2 Seeing the future: http://newvistadesign.net/dnlds/NUHS%20

Seeing%20the%20Future.pdf.

3   As with many wall-display quotes, there's no proof that Isaac Asmiov actually said this.

4   Again, this is broadly attributed to Winston Churchill, but nowhere to be seen in his collected writings.

5   Angela Davis in conversation with Jude Kelly at London's Southbank Centre, 11 March 2017.

6   Jerome Bruner makes a brilliant point about this, which Daisy Christodoulou cites in her *Seven Myths*: 'You cannot consider education without taking into account how culture gets passed on . . . it seems equally unlikely, given the nature of man's dependency as a creature, that this long period of dependency characteristic of our species was designed entirely for the most inefficient technique possible for regaining what has been gathered over a long period of time, i.e. discovery.'

7   Jonah Lehrer, 'A Physicist Solves the City', *New York Times*, 28 December 2010.

8   Geoffrey West, 'Scaling: The surprising mathematics of life and civilization', *Medium*, October 2014, at https://medium.com/sfi-30-foundations-frontiers/scaling-the-surprising-mathematics-of-life-and-civilization-49ee18640a8.

9   Luís M. A. Bettencourt, José Lobo, Dirk Helbing, Christian Kühnert, and Geoffrey B. West, 'Growth, innovation, scaling, and the pace of life in cities', PNAS, 24 April, 2007, vol. 104 no. 17.

10 Cited in Lehrer, *New York Times*.

11 Cited in Lehrer, *New York Times*.

12 The Harvard and World Bank academic Lant Pritchett has a brilliant passage on this topic in the opening of his book, *The Rebirth of Education*. Read it. He explains that education systems are particularly adept at using one of nature's great tricks, isomorphic mimicry, when an organism evolves to strongly resemble another in order to benefit. Our systems of learning do the same, borrowing governance and management practices and

implementing them at great expense for no measurable improvement in the learning of any children.

13 'Could do better: Bridge International Academies gets high marks for ambition but its business model is still unproven', *The Economist*, 28 January 2017.

14 https://www.theguardian.com/global-development/2017/may/05/beyond-justification-teachers-decry-uk-backing-private-schools-africa-bridge-international-academies-kenya-lawsuit.

15 Atul Gawande, 'Big Med', *New Yorker*, 13 August 2012.

16 Ernest Cline, *Ready Player One*, London, Arrow, 2012.

17 Todd Rose, *The End of Average: How We Succeed in a World That Values Sameness*, London, Penguin Books, 2017, p. 000.

18 J. K. Rowling, *Harry Potter and the Deathly Hallows*, Bloomsbury, 2007.

19 Alan Turing, 'Intelligent Machinery: A Heretical Theory', 1948.

20 Gideon Lewis Kraus, 'The Great A.I. Awakening', *New York Times*, 14 December 2016.

21 Ibid.

22 Anne-Marie Slaughter, 'Discard Old Ideas of a Leader of the Free World', *Financial Times*, 17 June 2017.

Afterword

1   This is taken from a talk Pharrell gave at Innovation Uncensored in New York in 2013. Pharrell's mom is a lifelong educator and he has much of interest to say on the topic. At an NYU Commencement speech he also told graduates, 'I like to say that I am forever a student.'

2   This speech is delivered in season 1, episode 18 of *The West Wing*.

3   UNESCO Institute for Statistics, *More Than Half of Children and Adolescents Are Not Learning Worldwide*, Fact Sheet No. 46, September 2017 (accessed at http://uis.unesco.org/sites/default/files/documents/fs46-more-than-half-children-not-learning-en-2017.pdf).

4 Kuan Chung, seventh century BC.

5 *Learning to fulfil education's potential*, 2018 World Development Report.

6 Saunders, *Braindead Megaphone*, p. 55.

7 Praveetha Patalay and Emla Fitzsimons, 'Mental ill-health among children of the new century: trends across childhood with a focus on age 14'. September 2017. Centre for Longitudinal Studies: London.

8 See for example, 'Investing in Mental Health', a 2003 report by the World Health Organization, at http://www.who.int/mental_health/media/investing_mnh.pdf. In 2017 the WHO also published 'Depression and Other Common Mental Disorders', which estimated that 300 million people worldwide suffer from depression today.

9 Kate Pickett and Richard G., *The Spirit Level: Why More Equal Societies Almost Always Do Better*, London, Penguin, 2009.

10 http://www.unesco.org/new/en/media-services/single-view/news/close_to_69_million_new_teachers_needed_to_reach_2030_educat/.

11 Amel Karboul, 'The global learning crisis and what to do about it', TED talk, October 2017.

INDEX